FRUIT PROCESSING INDUSTRY IN INDIA

Fruit Processing Industry in India

With Special Reference to the North-East

By

A.S. Yarso

AKANSHA PUBLISHING HOUSE

NEW DELHI-110 002(INDIA)

AKANSHA PUBLISHING HOUSE
4649B/21, Ansari Road
Daryaganj, New Delhi - 110 002
Phones: 011-23263193, 9811582579
Email: ektabooks@yahoo.com

Fruit Processing Industry in India

First Published, 2005

ISBN 81-8370-029-2

PRINTED IN INDIA

Published by M.P. Misra for Akansha Publishing House, New Delhi-110059 and Printed at Tarun Offset Press, Delhi.

Contents

	Foreword	vii
	Acknowledgement	ix
1.	Introduction	1
2.	Present Status of Fruit Processing Units	48
3.	Production and Marketing of Processed Fruit Products	63
4.	Performance Evaluation of Fruit Processing Units	93
5.	Finding of the Study and Suggestions	130
	Bibliography	139
	Abbreviations	154
	Appendices	157
	Index	190

Foreword

Despite the fact that India is the world's largest producer of fruits and second largest producer of vegetables, next only to China, with the annual production of 40 million tones of fruits and 60 million tones of vegetables, the country's share in the world trade in respect of processed fruits and food is only about one per cent. In this regard it is heartening to note that the Ministry of Food Processing, Government of India has targeted to increase India's share in the world trade of processed fruits to at least three per cent by the end of 2005. Unfortunately the fruit processing sector particularly in the north-eastern region of India is plagued with the problem of losses and wastage due to lack of adequate processing and storage facilities. It has been observed that in spite of having five food parks in the north-eastern region, the progress has not been satisfactory except in the state of Manipur and Assam. In this book Dr. A.S. Yarso has evaluated the performance of fruit processing units in both these states.

It gives me immense pleasure to write the foreword to this book **Fruit Processing Industry in India with Special Reference to the North-East**. Dr. A.S. Yarso, the author has done his doctoral dissertation on this topic under the supervision of Dr. A. Rajmani Singha, Department of Commerce, Manipur University. This book is the modified version of thesis. I comment both Dr. Yarso and his supervisor, Dr. A. Rajmani Singha for this much relevant study and now the same is presented in a book form for a wider readership.

—C.J. Thomas
Acting Director
ICSSR-NERC

April 2005
Shillong

Acknowledgement

This book is the outcome of my doctoral research work on **Performance Evaluating of Fruit Processing Industries in North-East India: A Case Study of Fruit Processing Units in Manipur and Assam.**

At the outset I am grateful to Dr. A. Rajmani Singha, Assistant Processor (Senior Grade), Department of Commerce, Manipur University for his valuable guidance during the course of my research work under his supervision. I put on record the valuable help and assistance received from the following officials: Mr S. Joykumar Singh, Deputy Director, FPI Section, Imphal; Mr. A.K. Gupta, Deputy Director, (F&VP) Eastern Region, Kolkata, Dr. Amiya Sharma, Deputy General Manager NEDFi, Guwahati; Professor S.V. Job, Pastor, LEF Church, Imphal; Professor Amalesh Banerjee, Vice-President, India Economics Association; Dr. S.B. Singh, Scientist, ICAR Shillong; Dr. G.P. Prasain, Associate Professor and Head, Department of Commerce, Manipur University and others such as: Dr. Tejamani Singh; Dr. Dhiren Meetei and Shri Ch. Ghanajit; Fr. Jose Palely, Principal, DBC, Maram; Dr. S.V. Ngachan, Joint Director, ICAR Imphal.

I would also thank the Proprietors and Managers of the Fruit Processing Units in Assam and Manipur and Dr. Tapan Chakravorty, Department of Statistics, Manipur University for his help in data analysis. I appreciate much ICSSR North Eastern Region Centre for their financial assistance under Continengy Grant for doctoral scholars and also Dr. C. J. Thomas for writing the foreword to this book.

I record my appreciation to Mr. M.P. Misra of Akansha

Publishing House for readily agreeing to publish this book and to Ms Christine Blah of ICSSR-NERC for her valuable secretrial assistance.

I would also like to record my heartfelt gratitude for the constant encouragement which I received from my family members particularly my sister, Mrs. A.S. Rachel, my uncle and aunt Mr. and Mrs. Chànce and my dear wife A.S. Ayao.

Lastly I am thankful to God Almighty for giving me the grace to complete my doctoral research and also helping me to bring out the revised thesis in a book form.

—A.S. Yarso

1

Introduction

Food processing industry is a part of fast moving consumer goods (FMCG) sector. 'The broad categories, that fall under food processing, include fruits and vegetables processing, food grain milling, dairy products, sugar, beverages like tea, coffee and cocoa products like chocolates, confectioneries, edible oils, vanaspati, tobacco and tobacco based products, processing and refrigeration of poultry and eggs, meat and meat products, processing of fish, fishing and fisheries beyond territorial waters, specialized packaging for food processing industries, beer, alcoholic drinks from molasses base, aerated waters, soft drinks and other processed food'.[1]

'In India, food processing industry is broadly classified into three groups, i.e. the primary food processing, the unorganised and cottage industries and processed food industries. The primary food processing constitutes rice hullers, flour chakkis, dal and oil mills. In the organised sector, there are over 820 flour mills, 409 fish processing units, 4,932 fruits & vegetables processing units, 114 meat processing units, 138 sweetened and aerated water (soft drinks) units, 90 milk product units, 429 sugar mills, 725 solvent extraction units and several other food processing factories'[2]. The unorganised and cottage industries consists of the various activities which comes under both the primary and the organised sector but smaller in the scale of operation comparing to the primary and the organised sector.

At present, 'food processing industry's size in India is estimated to be around Rs. 77,000 crore, of which, more than 42 per cent are in unorganised sector. Of the food processing industries, the oils and fats contribute 36 per cent, followed by dairy products 18.9 per cent, cold beverages 15.6 per cent, confectionary 3.5 per cent, processed fruits and vegetables 1.6 per cent and others 24.4 per cent. Since 1991, the growth of processed food units has increased significantly i.e. 3,925 units in 1991 are increased to 4270 units in 1995',[3] with a composite growth rate of 8.79 per cent during five years period.

'Looking into the Indian food habits, some companies are moving away from primary food processing to value added products like fruit juices, dehydrated foods, quick frozen food, and so on. A wide range of fruit juices from the fruits like guava and papaya has also been introduced recently by some of the reputed companies. Similarly, concentrates of guava and papaya have also been showing acceptance in export basket. Further, a few companies are also experimenting on to blend new fruit beverages such as of watermelon, litchi and sapota pulp etc.

In 1995-96, India exported pickles and chutney worth of Rs. 52.6 crore to some European countries. Due to relatively high profit margin world wide, the food processing industry has been accorded "Sunrise Industry" status by industrialists and policy makers. Although, 'the food processing industrial units have attracted Rs. 61,427 crore private investment in the first five years of reforms, the industry has failed to register the desired level of growth due to various short coming. To overcome some of these problems, in the 9th plan, an investment of around Rs. 10,000 crore has been proposed particularly in the areas of (i) processing of fruits and vegetables (ii) infrastructure development (iii) manpower development and (iv) research and development (R&D) studies'[4].

'The food processing industry is growing at an average rate of 20 per cent annually. The industry rank first in terms of employment generation with 19 per cent of the entire industrial labour force engaged in the agro food processing industry, ranks fifth in the country in terms of value addition' [5], and this industry (both organised and unorganised) has been identified by the government as a thrust sector mainly on account of high employment generation avenues and increase in value added output. According to D.P. Tripathi [6], "... as the food processing sector has a great employment generation possibility, it is expected that every Rs. 1000 crore investment will generate 54,000 person direct employment."

Technology

India is the world's largest producer of fruits and second largest producer in case of vegetables, next only to China, with an 'annual production of 40 million tonnes of fruits and 60 million tonnes of vegetables. It accounts for 1 per cent of its produce worth Rs. 23,000 crore is wasted every year due to lack of storage chains. At present, the existing cold storage capacity is estimated at about 26 million cubic metres, of which about 85 per cent of the facility is used for storage of potatoes'.[7]

In India 'one per cent of the total horticulture produce is converted into value added products and only 2 per cent of the total fruits and vegetables produced in the country are processed as against 40 per cent in developing countries and 70 per cent in developed countries'[8], which is the indication of poor expl(tation in the field of horticultural industries. According to S.R. Prusty,[9] "only modern technology, quick transport and better storage facility can enable the industry to grow to its full potential. In order to adopt greater efficiency, the industry needs to infuse latest technology that would provide economies of scale and cost effectiveness, greater use of radiation technology in processing and application of biotechnology will help in processing and improving varieties of fruits and vegetables. Thus, by taking appropriate technological measures India can still progress in fruit and vegetable processing industries to a commendable position.

Union Budget 1999-2000 and 2000-01

In the Union Budget 1999-2000, the Government had 'reduced the excise duty by 2 per cent (from 18 per cent to 16 per cent) on chocolates, malted food drinks, instant coffee and sugar confectionery. The impact of the budget on this sector was positive mainly on account of the initiatives outlined in the budget. The budget had given priority sector status to the food processing industry for the purpose of bank lending. The budget aimed at improving cold distribution chains and reduction in wastage' [10].

The Union Budget 2000-01 has highlighted the importance of the food processing industry. The 'measures taken by the Finance Ministry for the benefit of the industry include : (i) specified cold chain equipment exempted from excise duty (ii) import duty on agricultural produce (other than cereals) and marine products gone up to 35 per cent plus 10 percent surcharge from the existing rate of 5 to 15 per cent. However, duties on products like biscuits, dairy products and bread and cakes have been increase from 8 per cent to 16 per cent'[11]. From the above two annual budgets, it has clearly shown the tax structure for food processing industries. India has one of the highest tax rate charged/against the food processing industries as compared to other Asian countries. It is the only country to levy excise duty on processed foods apart from Thailand.

Foreign Direct Investment

India, followed a liberal policy since 1991 for foreign direct investment (FDI) in food-processing sector. Post harvest technology, infrastructure building, cold chains and packaging are some of the thrust areas in which the country is seeking competent overseas collaborations.

Food processing sector, considered to be one of the "sunrise" industries is speedily increasing the share of FDI over the years. 'Till November 1999, 5,718 Industrial Entrepreneurial Memoranda (IEMs) have been filed in the food processing sector envisaging an investment of Rs. 53,700 crore. Apart from these,

1,111 approvals have been granted till October 1999 envisaging an investment of Rs. 19,086 crore for setting up of 100 per cent EOUs/industrial licences, in various sectors of food processing out of the total investment of Rs. 72, 786 crore approved in the sector, the foreign investment involved is Rs. 9,125 crore. 848 projects have already gone into commercial production till September 1999'[12].

The present study is related to fruits processing industry, hence our subsequent discussions will be confined to fruits processing sector only. Of which fruit processing trend in India as a whole is highlighted and emphasis is given more on fruit processing in North East India.

Fruit Crops in India

Fruits not only improve the quality of the diet but also provide essential ingredients like vitamins, minerals, carbohydrates, dietary fibre etc. They also exhibit a high potential for generating employment in rural areas. Because of high productivity and value, fruit crops in India provide much better economic returns per unit area compared to cereal crops and hence it is a good source of foreign exchange earnings for the country.

Table 1.1

Production of fruits crops in India

(million tonnes)

Year	*Fruits*
1991 -1992	28. 63
1992 - 1993	32.96
1993 - 1994	39.47
1994 - 1995	38.60
1995 - 1996	41.51
1996 - 1997	46.97

Source : Economic Survey of India, 1997-98

India is the largest producer of fruits in the world. The production of fruits in the country has tripled after independence. The production figures for the last few years are given in Table-1.1. 'Presently the area under fruits cultivation is estimated to be 3.94 million hectare'[13].

Though the production figures are comparatively high a significant amount of qualitative and quantitative losses occur from the period of harvest till consumption. 'The magnitude of post harvest losses estimated on fresh fruits range between 5 to 25 per cent in developed countries and between 20 to 25 per cent in developing countries. Several studies had indicate that in India the loss occured is estimated to be 22 to 40 per cent, which amounts to about Rs. 30,000 - 40,000 million annually'[14].

Taking the population of the country as 1,027 million (as per 2001 census) and the minimum per capita requirement of fruits to be 85g per day, the net requirement will be 31.86 million tonnes. At present, the net availability (by taking a post harvest loss of 40 per cent) of fruits is 26.18 million tonnes. Thus, it shows that though we have gained a significant jump in production but due to lack of proper post harvest measures it created a shortfall of 5.68 million tonnes for fruits. In order to bridge this gap and also for processing and export, emphasis need to be given for reducing the losses and increasing the production and productivity of horticultural products. As it has been reported that fruits account for 65.8 per cent of total area and 90.2 per cent of all horticultural produce respectively in India. But it has been estimated that just around 2 per cent of fresh fruits are processed or preserved in India compared to 30 per cent in Thailand, 70 per cent each in Brazil and USA, 78 per cent in Philippines, 80 per cent in South Africa and 85 per cent in Malaysia[15]. Again, in India, not even 45 per cent of the installed capacity of fruit processing industry is utilized due to non-availability of raw materials at reasonable price with uniform physio - chemical quality parameters.

According to H.P. Singh[16], the low productivity of major fruits, excepting grapes compared to other competing countries, inadequate availability of raw materials of desired quality at reasonable price, outdated processing technology, high cost and poor quality of packaging material, multi-level taxes and duties etc. are the major constraints, being continually confronted by the fruits processing industry. Unless, these bottlenecks are lessened considerably, neither the domestic demand nor exports could be met properly from this sector.

The North East Profile : Following are the brief profile of North East India covering geographical location, physical features, climate, area and population. These are highlighted below :

Geographical Position

The North Eastern Region of India comprises of 7 states namely Arunachal Pradesh, Assam, Manipur, Meghalaya, Mizoram, Nagaland and Tripura. The region lies between 21°57′ degree to 29°28′ north latitude and 89°40′ degree to 97°255′ east longitude. The total geographical area of the region is 2.55 lakh sq. km., which is about 8 per cent of the country's total area. The region is connected with the rest of the country through West Bengal by a narrow corridor of 56 km, which runs below the foothills of Bhutan and Sikkim. It has 1350 km long border with China in the North, 900 km long border with Myanmar in the East. The boundary with rest of India is only 38 km[17], which is the width of the narrow corridor connecting the region with West Bengal.

Physical Features

The physiography of the region is divided into three divisions, namely Meghalaya Plateau, the North Eastern Hills and Basin and the Brahmaputra Valley. The North Eastern Hills and Basin alone account for 65 per cent of the total land area while the Brahmaputra Valley and the Meghalaya Plateau cover 22 per cent and 13 per cent of the area respectively[18].

Climate

The climate of the region varies from cold below freezing

temperatures in the upper Himalayan stretches of Arunachal Pradesh to more than 30 degree centigrade temperature in Guwahati in August. Within each state again the climatic conditions vary considerably giving the region an exciting mix of agro-climatic possibilities. The rainy season continues from mid March to middle of October.

The forest land in North East covered more than 50 per cent of the land area. Over all 42 per cent is classified as dense forest and 58 per cent as open forest. Around 33 per cent of the land is under jhum cultivation.

Area and Population

The geographical area of the states, total population, percentage of urban population and density of population are presented in Table -1.2.

Nearly 85 per cent of the total population are rural and agriculture is the most dominant vocation of the people. The contribution of agriculture to the state domestic income is much higher in this region, except for Arunachal Pradesh, Meghalaya and Nagaland, mining in case of Meghalaya and forestry and logging in case of Arunachal Pradesh and Nagaland are important contributors to state domestic product. Out of the constituent states Assam accounts for nearly 70 per cent of the human population with the highest density of 339 persons per sq. km. Tripura is next with the density of population having 304 persons per sq. km. While the rest of the states are rather thinly populated, Arunachal Pradesh having only 13 persons per sq. km. The region has a provisional total population of 384.95 lakh as per 2001 census shown in the Table -1.2. Of which only around 15 per cent are urban and more than 70 per cent of the total population are depending directly on agriculture, hence the entire region is based on rural economy.

Despite massive development effort made by the central and states governments through various five year plans, the problem of mass poverty, growing unemployment and large scale migration to urban area persist in rural areas. Unless, the

Table - 1.2

Area and Population of North Eastern Region

STATE	Geography area (sq. km.)	Population as per 2001 census			% Urban Population to total population	Density of Population / Persons per sq. km. (2001)
		Total	Rural	Urban		
Arunachal Pradesh	83,578	1,091,117	868,429	222,688	20.4	13
Assam	78,523	26,638,407	23,248,994	3,389,413	12.7	339
Manipur	22,356	2,388,634	1,818,224	570,410	23.8	106
Meghalaya	22,489	2,306,069	1,853,457	452,612	19.6	102
Mizoram	21,087	891,058	450,018	441,040	49.5	42
Nagaland	16,527	1,988,636	1,635,815	352,821	17.7	120
Tripura	10,477	3,191,168	2,648,074	543,094	17.0	304
Total	255,037	38,495,089	32,523,011	5,972,078		

Source : Census of India, 2001; Statistical Organisation India.

rural economy is improved, the burden of poverty cannot be lessened nor the problem of unemployment could be solved. About 37 per cent of the rural population still lives below poverty line. This is because of the landless agricultural labourers (about 55 lakh house holds) who do not have sufficient income from agriculture. However development of agriculture alone can not improve the over all development of the rural economy, there is a need to diversify the rural economy by starting more viable and suitable small scale and cottage industries like fruit and vegetable processing units and other horticultural industries in the rural areas. The establishment of such industries will provide increased alternative employment especially to the rural landless and small farmers. To boost up the development of this sector, the role of both the state and central government are highly essential.

PART - II

Role of the Government in Fruit Processing Industry

The role of the government, both state and central, for the development of food processing industries in general and fruits processing industry in particular is very significant. Infrastructural development, incentives and subsidies for transport and other benefits for the development of small scale industries can not be met by the private investment. It is only through government initiative, policy and plans that development of industries can be achieved if not fully. We shall discuss the role of the governments in formulating and implementing the policies for the development of food processing industries and new industrial policy and other concession for the North Eastern Region in the following pages.

Ministry of Food Processing Industries (MFPI)

The Ministry is under the Ministry of Agriculture, Government of India with the vision to motivate farmers and food processors and to provide an interactive coupling between technology, economy, environment and society for speedy development of food processing industries, to build up a substantial base for production of value added agro-food products both for domestic and export markets, a strong emphasis on food safety and quality enabling the farmers especially to realise direct benefits of new technology and marketing network and to ensure adequate availability of quality food products for the consumer at economic prices.

The scope of the Ministry includes development of fruit and promotion of food, grain milling including canning and freezing and technical assistance to the industry also form a very important part of its activity. In addition planning and developing of industries relating to bread, oil seeds, breakfast food, biscuits, confectionery, specialised packaging, including non alcoholic beer, aerated drinks also fall within the scope of this ministry.

MFPI's Role in Fruit Processing Sector

The policy and schemes of the Ministry can better explain the central government's role towards the development of fruit processing sector in India.

The Policy

The government has laid down the following policies for the development of food processing industries which includes fruit processing sector in India :-

1) Creating Enabling Environment

The policy will seek to create an appropriate environment for the entrepreneurs to set up food processing industries through:

- Fiscal initiative/interventions like rationalization of tax structure on fresh foods as well as processed foods and machinery used for the production of processed foods.

- Harmonization and simplification of food laws by an appropriate enactment to cover all provisions relating to food products so that the existing system of multiple laws are replaced and also covering issues concerning standards nutrition, advantageous goods, future marketing, equalization fund etc.
- A concerned promotion campaign to create market for processed foods assistance to Industry Associations, NGO's/ Cooperatives, Private Sector units, State Government Organization for undertaking generic market promotion.
- Efforts to expand the availability of the right kind and quantity of raw material round the year, increasing production and improving productivity.
- Strengthening of database and market intelligence system through studies and surveys.
- Strengthening extension services to the farmers and cooperatives in the areas of post harvest management of agro-produce to encourage creation of preprocessing facilities near the farms like washing, fumigation, packing etc.
- Efforts to encourage setting up of agro-processing facilities as close to the area of production as possible to avoid wastage and reduce transportation cost.
- Promotion of investments, both foreign and domestic.
- Simplification of documentation and procedures under taxation laws to avoid unnecessary harassment arising out of mere technicalities.

2) Infrastructural Development

The policy will facilitate for the development of infrastructure through :

- Establishment of cold chain, low cost precooling

facilities near farms, cold storage and grading, sorting, packing facilities to reduce wastage, improve quality and shelf life of products.

- Application of biotechnology, remote sensing technology, energy saving technology and technologies for environmental protection.
- Building up a strong infrastructural base for production of value added product with special emphasis on food safety and quality matching for international standards.
- Development of packaging technologies for individual productions, especially cut-fruits and vegetables, so as to increase their shelf life and improve consumer acceptance both in the domestic and international markets.
- Development of new technologies in food processing and packaging and also to provide for the mechanisms to facilitate quick transfer of technologies to field through a net work of Research and Development (R&D) institutions having a central institute at the national level with satellite institutions located strategically in various regions to cover up the whole country and to make available the required testing facilities. This could be done by establishing a new institution or strengthening an existing one.
- Development of area specific agro food parks dedicated to processing of the predominant produce of the area e.g. apple in Jammu and Kashmir, litchi in Bihar, mango in Maharashtra and Andhra Pradesh etc.
- Development of anchor industrial centre and/or linkage with anchor industrial units having network of small processing units.
- Development of agro-industrial multi products units

capable of processing a cluster of trans-seasonal produces.

3) *Backward Linkage*

The policy will promote for the backward linkages of food processing industries through:

- Establishing of a sustained and lasting linkage between the farmers and the processors based on mutual trust and benefits by utilizing the existing infrastructure of cooperative, village panchayats and such others institutions.
- Development of future market in the vested interest of both the farmers and the processors ensuring a minimum price stability to the farmer and sustained supply of raw material to the processor.
- Mechanism to reduce the gap between the farm gate price of agro-produce and the final price paid by the consumer.
- Setting up of an equalization fund to ensure sustained supply of raw material at a particular price level and at the same time to plough back the savings occurring in the eventuality of lower price to make the fund self- regenerative.

4) *Forward Linkage*

The policy will promote forward linkage through:

- Establishment of a strong linkage between the processor and the market to effect cost economies by elimination of avoidable intermediaries.
- Establishment of marketing network with an apex body to ensure proper marketing of processed products.

- Development of marketing capabilities both with regard to infrastructure and quality in order to promote competitive capabilities to face not only the WTO challenges but to undertake exports in a commendable form.

5) Very Special Provision

The following shall receive higher priority and special consideration in policy and plans. The hills areas, islands and the Integrated Tribal Development Programme (ITDP) areas in the country to be given not only special attention but also special consideration. The fiscal incentives like excise duty/ sales tax concession and the tax holidays to be provided not only to those units which are set up in these areas but also to these units which though set up outside these areas near the market centre, are engaged in processing the produce coming from these areas. Tax holiday for food processing units with the exception of liquor, cigarettes and aerated drinks and similar luxury items, for a period of 10 years.

Research and Development Schemes

The Ministry also provides assistance for research and development projects in fruits and vegetables processing sector. So far, assistance has been provided for the following research and development project:

1. Centre for Technology Development, Bangalore for setting up of an Analytical and Quality Control laboratory.
2. APEDA, New Delhi for vapour heat treatment facilities for mangoes and other fruits for exports; and
3. M/S Fresh Marketing Cooperative Society Ltd. Hyderabad for installation of a Solar Refrigeration unit received from FAO.

Ministry in collaboration with the department of

Biotechnology is setting up Biotechnology Research Centre for national facility for food safety at CFTRI, Mysore at an estimated cost of Rs. 373.65 lakh. For this project, assistance to the tune of Rs 100 lakh has already been provided in 1994-95 and 1995-96. Also assistance has been provided for setting up of post harvest technology institute in Pune district in Maharashtra.

Achievements of MFPI with reference to North Eastern Region

Since the researcher is of the North East India, it will be more appropriate to mention the policies and schemes of MFPI which were implemented in the region.

The plan assistance provided to the fruit processing sector of the region are given below :

1. During 1995-96 assistance has been provided for setting up of 54 food processing and training centre in nine states of India of which Assam is one from North East India.

2. Under the scheme for providing assistance for creation of infrastructural facilities for fruit processing including mushroom processing, hops and setting up of industrial estates in the first 3 years of the 8th five year plan, assistance has been provided for a number of project received from 13 states of India of which Nagaland is one from North East India.

3. During the year 1995-96, assistance has been provided to one cooperative society in Nagaland for setting up of mushroom infrastructure and processing unit.

4. Under the scheme for setting up/expansion/ upgradation of fruit processing units and backward linkage between the processors and the farmers etc., assistance is provided to state public sector undertakings from North Eastern States and 3 other states of the country.

5. Under the scheme for setting up/expansion/upgradation, mentioned above, another sub-scheme has been formulated for being operated during the remaining 2 years of the 8th five year plan i.e. 1995-96 and 1996-97 for providing assistance for setting up of small and tiny units of food processing industries in the unorganised sector. So far assistance has been released in a number of projects received from 13 states of India of which Assam, Nagaland and Mizoram are amongst the states.
6. For strengthening of backward linkages and encouraging contract farming so far as many as 10 processing units have been provided assistance in 7 different states of India of which Assam is one. These fruit processing units have adopted contract farming. It is reported that more than 3,500 farmers have been brought under the umbrella of contract farming.
7. During the period of review, the department has granted assistance to 13 different units of varying amounts for promotion of food processing industries and development of infrastructure facilities of which 3 are from the North Eastern States they are :-

   i. During the period under review, the department has provided a grant of Rs. 42.33 lakh to M/S Rishang Keishing Foundation for Management of Tribal Area (MATA), Manipur for setting up of a ginger processing plant in Ukhrul district of Manipur. The unit has started production of ginger oil from January, 2002.

   ii. A grant-in-aid of Rs. 5.72 lakh was provided to Jonthra Multi purpose Cooperative Society Ltd., Nagaland for setting up of a fruit and vegetable processing unit in Mon district, in Nagaland.

   iii. During the period under review M/S Integrated Economic Development Society Manipur was provided a grant-in-aid of Rs. 4.36 lakh by the department for setting up of fruits and vegetables dehydration unit in Imphal district in the state of Manipur.

Features of the New Industrial Policy and other Concession for the North Eastern Region

The achievements of MFPI discussed above are few comparing to the vast potentialities of horticultural industries in the North Eastern Region.

In view of the continuing backwardness of North Eastern Region the need for a new and synergetic incentive package was widely felt to stimulate development of industries. Hon'ble Prime Minister of India during his visit to North East Region in October, 1996 assured that a separate industrial policy would be announced for the industrial development of the North Eastern Region for which expert group committees were constituted by the Ministry of Industry and Planning Commission. Based on the recommendations and proposals finalised by these expert groups/committees, the Government of India approved the new Industrial Policy and other concessions in the North Eastern Region.

A: Development of Industrial Infrastructure

1) Growth Centre

Currently the funding pattern of the growth centres envisaged a central assistance of Rs. 10 crore for each growth centre and balance amount to be raised by the state government has approved that entire expenditure on the growth centres would be provided as central assistance, subject to a ceiling of Rs. 15 crore.

2) Integrated Infrastructure Development (IID) Centre

In respect of the IID centres, the funding pattern would be

changed from 2:3 between government of India and SIDBI (Small Industrial Development Bank of India) to 4:1 and the funds from the government of India would be in the form of grant.

B: Transport Subsidy Scheme

The transport subsidy scheme will be extended further in so far as North Eastern States are concerned, for a period of another 7 years i.e. upto 31st March, 2007 being continuous with the tenth five year planning on same terms and conditions as are applicable now.

C: Fiscal Incentives to New Industrial Units and their Substantial Expansion

1) Total Tax Free Zone

Government has approved for converting the growth centers and IIDs units to a total tax free zone for the next 10 years. All industrial activity in these zones would be free from income tax, excise for a period of 10 years from the commencement of production. State government would be requested to grant exemptions in respect of sales tax and municipal tax.

2) Capital Investment Subsidy

Industries located in the growth centre would also be given capital investment subsidy at the rate of 15 per cent of their investment in plant and machinery, subject to a maximum ceiling of Rs. 30 Lakh. The Commercial Banks and the North East Development Financial Corporation (NEDFC) will have dedicated branches/counters to process applications to term loans and working capital in these centres. While sanctioning assistance NEDFi and Commercial Banks would take a liberal view of the debt equity ratio.

3) Interest Subsidy on Working Capital Loan

An interest subsidy of 3 per cent of working capital loan would be provided for 10 years after the commercial production. The working capital requirements would worked out as per the Nayak Committee.

4) Excise Benefits

Government of India vide its gazette notifications No. 32/99 dated 8-7-99, No 33/99 and NO 48/99 dated 8-8-99 (annexure-F) has given sweeping concessions on excise duty. All excisable duty produced in the factory located in the growth centres, IIDs etc. in the state have been exempted from payment of excise duty. Goods produced in specified industries located in areas outside the growth centres/IIDs etc. have also been exempted from payment of excise duty. It has been decided that while the manufacturer will not be required to pay excise duty on their production, the user of such products, using them as inputs will get the MODVAT (Modified value added tax) credit. At the same time, the manufacturers are being permitted to avail of MODVAT credit only paid input material received by them for their finished products.

The above excise exemption shall apply only to the following kind of units namely:

i) New industrial units which have commenced their commercial production on or after 24/12/97.

ii) Industrial units existing before 24/12/2002 but which have undertaken substantial expansion by way of increase in install capacity by not less than 2.5 per cent on or after 24/12/97.

Further the above exemption shall apply to any of the aforesaid units for a period not exceeding 10 years from the date of publication of the above notifications or from the date of commencement of commercial production whichever is later.

D: Relaxation of PMRY Norms

The PMRY would be expanding its scope covering areas of horticulture, piggery, poultry, fishing small tea gardens etc. so as to cover all economically viable activities. PMRY would have a family income ceiling of Rs. 40,000 per annum for each beneficiary with his/her spouse and upper age limit will be relaxed up to 40 years. Projects costing upto Rs. 2 lakhs in other than business sector will be eligible for assistance. No collateral will be insisted for projects costing upto Rs. 1 lakh, group financing upto Rs. 5 lakhs will be eligible. Scheme will have a subsidy component @15 per cent with an upper ceiling of Rs. 15,000. The margin money may vary from 5 per cent to 12.5 per cent of the project cost to make the subsidy and margin contributions at 20 per cent of the project cost. PMRY would continue to have entrepreneurship training component as per the existing rule.

E: Other Incentives Proposed

i) A comprehensive insurance scheme for industrial units in the North East India will be designed in consultation with General Insurance Corporation of India Ltd. and 100 per cent premium for a period of 10 years would be subsidized by the central government.

ii) A one time grant of Rs. 20 crore will be provided to the North East Development Financial Corporation (NEDFi) by the central government through NEC to fund Techno-Economic studies for industries and infrastructure which are most viable and best suited to this region.

iii) State government may consider setting up a "Debt purchase window" by the NEDFi which buys the debt of the manufacturing units particularly in respect of the supplies made to the government departments so as to reduce the problem of blocking funds for these units.

iv) For the developments of markets in North East specially of export of products of North East to the neighboring countries particularly, Bangladesh, Myanmar and Bhutan and South East Asian Countries in general would be explored.

v) It may be considered to provide assistance for restructuring State PSUs (Public Sector Units) from National Renewal Fund.

vi) The community pattern of land holding in large areas of North Eastern Region does not lend himself to providing collateral security, under conventional bank lending. RBI has constituted a committee to look into this issue. An appropriate system of "guarantees" will involved for North Eastern Region.

F: Procedure for Release of Assistance Under the New Initiatives

It is approved that the transport subsidy budget may be released by a designated agency on the basis of the recommendation of the SLC (State level committee). It is proposed that NEDFi may be designated as the nodal agency for release of transport subsidy in North Eastern States. NEDFi may be paid administrative expenses for this service which may be decided in consultation with IDBI.

G: Development of Village and Small Industries (VSI) Sector

Weavers Service Centers (WSCs) in North Eastern Region and Indian Institute of Handloom Technology (IIHT) at Guwahati would be suitably strengthened to provide technology and training support to the weavers. National Handloom Development Corporation (NHDC) will give priority in supply of hank yarn to the North Eastern Region. All the four variety of silk would be covered under the Mill Gate Price Scheme. Priority would be given to the North Eastern Region in scheme of setting up market complexes and permanent exhibition facilities. A new design centre for development of handicraft would be set up in North Eastern Region. To upgrade the skill of artisan, advance training programme would be organised. New emporia will be set up and financial assistance for renovation of existing emporia would be provided. The central silk board will give priority to North Eastern Region in implementation of its scheme.

Tentative List of Horticulture Based Industries Appropriate for Development in the North Eastern Region

1) Fruit and Vegetable Processing Industries

a. Canned/ Bottled products

b. Aseptic packaged products

c. Frozen products

d. Dehydrated products

e. Oleoresin

2) Consumer Industry

a. Non-Alcoholic beverages

3) Food Packaging Industries

4) Horticulture Industry

Technology Mission on Horticulture Development of North Eastern States

1. Horticulture in North Eastern States

The North Eastern (N.E) Region accounts for nearly 8 per cent of the country's geographical area and about 4 per cent of the population of the country. The topography of the region ranges from hills and mountains to riverine plains and plateaus. The climatic condition of the region is diverse which varies from temperate to sub tropical and tropical which is suitable for horticulture crops. The horticulture sector, which includes fruits, vegetables, spices, plantation crops, floriculture, medicinal and aromatic plants, cashewnut etc. has ample scope and potential for development which has remained unexploited. To harness the potential of horticulture in the region, Government of India is implementing 'Centrally sponsored scheme on Technology Mission for Integrated Development of Horticulture in North Eastern States of India including Sikkim'[19].

The Scheme was approved on 27th February, 2001 with an outlay of Rs. 229.38 crore. Implementation of the scheme started in 2001 - 2002 with an allocation of Rs. 6724.96 lakh.

2. ***Objectives of the scheme*** : The objectives of the Mission are:

i) to harness the potential of horticulture of the region by increasing production and productivity of horticultural crops.

ii) to maximize economic, ecological and social benefits through desirable diversification.

iii) to develop additional infrastructure for production of planting material, storage and processing of horticultural produce.

iv) to provide skilful employment in the region.

3. ***Structure of Mission***

The mission is being implemented in a mission-made approach through four Mini-Missions.

MINI MISSION - I : Research -coordinated and implemented by Indian Council of Agriculture Research. This Mini Mission concentrates on technology generation approach to the region. Major components under MMI are supply of basic seed and planting material, technology standardization, refinement on farm demonstration and training.

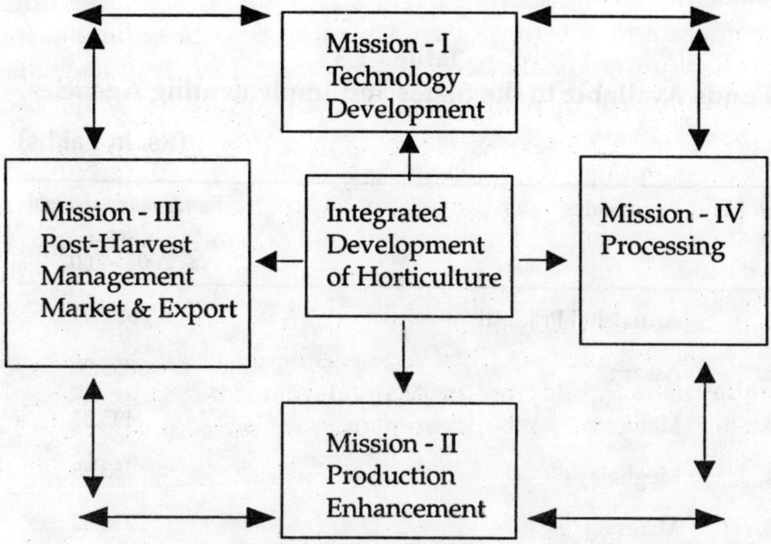

Figure - 1.1 Technology Mission for integrated Development of Horticulture in North Eastern State of India including Sikkim.

MINI MISSION-II :Productivity coordinated by the Department of Agriculture and Cooperation and implemented by the state Department of Horticulture and Agriculture. This missions aims at increasing production and productivity of horticulture crops by adoption of improved products technologies.

MINI MISSION-III : Post harvest management and marketing coordinated by Department of Agriculture & Cooperation implemented by Directorate of Market Intelligence, National Horticulture Board. This Mini Mission aims for efficient post harvest management, techniques, which include development of cold storage facilities, efficient transport and marketing facilities.

MINI MISSION IV : Processing coordinated and implemented by Ministry of Food Processing Industries, Government of India. This Mini Mission aims at promoting processing industry for value addition to horticultural produce by promoting new processing units, upgradation of existing units.

Table - 1.3

Funds Available to the States and Implementing Agencies.

(Rs. in Lakhs)

Sl. No.	*States*	*Funds made available during 2001 - 2002*
1.	Arunachal Pradesh	957.69
2.	Assam	1055.90
3.	Manipur	495.31
4.	Meghalaya	753.59
5.	Mizoram	735.82
6.	Nagaland	703.01
7.	Sikkim	744.25
8.	Tripura	749.68
9.	Space Application centre (ISRO)	25.00
10.	National Informatic centre (NIC)	22.84
11.	Infrastructure	7,688.00
12.	Undisbursed	419.40
13.	SPAC service charge	54.74
	Total	14,405.23

Source : Department of Agriculture & Cooperation, Govt. of India.

During the year 2001 - 2002, the above funds have been made available to the states and implementing agencies based on the action plan prepared by the respective states.

4. Selection of Beneficiaries

Beneficiaries at field level are identified by Nodal Agencies, Department of Horticulture/Agriculture concerned, as per requirements and priorities of the state.

5. Monitoring of Scheme

The implementation of scheme is being monitored through steering committees constituted at the central, state and mission levels.

i) Central Steering Committee : This committee is headed by Secretary (A & C) and meets once a year to oversee the progress of scheme and suggests measure for improving the implementation.

ii) Steering Committee for Mini Mission - I is headed by Director General, ICAR.

iii) Steering Committee for Mini Mission - II and Mini Mission - III are headed by Special Secretary, Department of Agriculture & Cooperation.

iv) Steering Committee for Mini Mission - IV is headed by Secretary, Department of Food Processing Industries.

v) Review Meetings are being held periodically by special Secretary.

vi) State Level Steering Committee : Each state has constituted a state level Steering Committee under the chairmanship of chief Secretary of the state. The Steering committee meeting is held at least once in a year to oversee the implementation of the mission in the states.

vii) An annual conference to review the progress of mission and finalise the action plan.

6. Awareness Campaigns cum Workshops

In order to make people of North East aware about the benefits of diversifying to horticulture crops and with a view to let public know about the launching of Technology Mission on Horticulture programmes of organising awareness campaigns for the people of the state from all walks of life are invited and mission objectives are explained in detailed. Such campaigns have been appreciated by all and have brought a new awakening that diversification to horticulture could reduce economic imbalances in the North Eastern States and stop Jhum cultivation altogether.

Institutional Support to Fruit Processing Industry

Following are some of the institutions and government agencies who are directly and indirectly involving for the development of fruit processing industry in particular and food processing industries in general. They are :-

1. Ministry of Food Processing Industries (MFPI)

To link the farmer, the processor, the consumer and the usher in the new era of prosperity, leads to the creation of Ministry of Food Processing Industries (MFPI) in India during 1988.

This Ministry is at present under the Ministry of Agriculture, Government of India. MFPI is responsible for developing a strong and vibrant food processing sector with more emphasis on stimulating demand for appropriate processed foods, achieving maximum value addition and by-product utilisation, creating increase of job opportunities particularly in rural area. It is also responsible for enabling farmers to reap the benefits of modern technology and creating surpluses for export.

In discharging the above responsibilities, the MFPI has the policy of supporting formulation and implementation of policies for food processing industries within the overall national priorities and objectives and facilitating the creation of a conducive environment for healthy growth of the food processing sector. It also help in promoting rationalization of tariffs and duties relating to food processing sector and harmonisation of Food Laws.

MFPI also plays a developmental role by giving assistance under various plan schemes and widening the Research and Development (R & D) base in food processing. The Ministry also gives assistance for setting up analytical and testing laboratories, active participation in the laying down of food standards and their harmonisation with the international standards and developing human resource to meet the growing requirement of managers, entrepreneurs and skilled workers in the food processing industry.

In order to create awareness about the potential and prospect of food processing industries in the country, this ministry provides assistance for organising workshops, seminars, surveys, exhibitions and fairs. To disseminate information on processed foods in India and abroad, it gives assistance for publications and films in the field of food processing. The Ministry also plays a regulatory role for implementation of Fruit Products Order (FPO), 1995.

2. *National Horticulture Board*

National Horticulture Board was set up in 1984 by the Government of India in order to increase production and marketing of horticultural products and to look after different aspects of the horticulture industry. The Board is an important instrument of the Government of India for the integrated development of horticulture in the country. The Union Agriculture Minister is the President of the Board and Union Agriculture Secretary is the Chairman of the Managing Committee of the Board. The main objectives of the National Horticulture Board are to:

a) encourage and promote the development of horticulture industry in the country;

b) assist the establishment of growers societies to advance the economic and social status of farmers;

c) assist and develop infrastructure for post-harvest technology, and

d) provide technological, financial and other assistance to various organisation for development of horticulture.

The National Horticulture Board has taken a lead in meeting all the objectives mandated to it and various programmes ranging from production through post harvest handling to marketing including export and processing have been initiated in the country for the integrated development of horticulture.

3. *Agricultural and Processed Food Products Export Development Authority (APEDA)*

Agricultural and Processed food products Export Development Authority (APEDA), came into existence under the Ministry of Commerce, Government of India in 1986. APEDA came into being in order to further develop India's agricultural commodities and processed foods, and to promote their exports. It aims at maximising foreign exchange earnings through increased agro exports, to provide better income to the farmers through higher unit value realisation and to create employment opportunities in rural areas by encouraging value added exports to farm produce.

To achieve the above objectives APEDA has initiated various development programmes for the growth and development of agricultural and processed food products, such as : development of data base on products, markets and services, organisation of product promotions abroad and visits of officials and trade delegations abroad, participation in international trade fairs in India and abroad, organisation of 'Buyer Seller Meets' and other business interactions, provides recommentory, advisory and support services to the trade and industry, problems solving in government agencies and organisations, RBI, Customs, import/export procedures, problems with importers through Indian Missions abroad.

APEDA offers financial assistance under various schemes which seek to promote and develop agro-exports. Financial assistance under these schemes are available to exporters, growers, trade associations, governmental agencies etc.

4. National Co-operation Development Corporation (NCDC)

The genesis of NCDC is to be traced to the report of the All India Rural Credit Survey Committee (1954), which advocated an integrated approach to co-operative development in the rural sector. As a sequel to the acceptance of the said committee's recommendations, the Reserve Bank of India Act was amended to enable it to provide substantial support to agriculture/ co-operation credit programme in the country. A new statutory body called the NCD & W Board was set up in 1956 to provide institutional support for development of co-operative marketing, processing, warehousing and storage. Consequently the subject "Warehousing" was transferred to the Ministry of Food Processing and a separate "Department of Co-operation" under the Ministry of Agriculture, a new body NCDC was created by an Act of parliament in 1962. NCDC has been conceived as a nodal organisation for meeting the post-harvest requirements of farmers through co-operative societies.

NCDC as per its statute is mainly concerned with promoting, developing and financing post harvest activities in agriculture sector. These comprise of marketing, storage and processing of agricultural produce besides supply of seeds, fertilizer and other agricultural inputs.

5. National Bank for Agriculture and Rural Development (NABARD)

The National Bank for Agriculture and Rural Development was set up on 12th July 1982, under an Act of parliament by initially merging the Agricultural Credit Department and Rural planning and Credit Cell of Reserve Bank of India (RBI) and the entire undertaking of Agricultural Refinance and Development Corporation (ARDC). The National Bank for Agricultural and Rural Development (NABARD) is an apex development bank of the country for supporting and promoting agriculture and rural development.

NABARD provides short-term, medium and long-term refinance facilities to the institutions for the following purposes:

a) Agricultural production, operations and marketing of crops by farmers and farmers' cooperatives societies etc.

b) Production and marketing activities of village & cottage industries, small scale and tiny industries.

c) Investment in agriculture and allied activities such as plantation/horticulture, setting up storage and market yards, agro-processing, agro-services centres etc.

d) Activities of voluntary agencies/self help groups for rural poor.

e) Investment in share capital/securities of institutions concerned with agriculture and rural development.

6. North Eastern Regional Agricultural Marketing Corporation Limited (NERAMAC)

NERAMAC was incorporated on 31st March 1982. It was promoted by the North Eastern Council, Shillong under the administration of Ministry of Food Processing Industries, Government of India, New Delhi.

Keeping in view its mandate for agricultural prosperity by undertaking development of marketing infrastructure, the corporation has been undertaking the following activities since its inception.

Export of processed fruit products of the SSI (Small Scale Industry) units of the region thereby ensuring capacity utilisation with commitment of marketing. Export of agri-inputs like true potato seeds has also created market for North Eastern origin products in the global market.

Apart from marketing of processed fruit products the corporation has undertaken marketing of fresh fruits and vegetables also. In addition to the above, the corporation is involved in marketing of vegetables, cashewnut, cereal products, spices and minor forest produces of the North Eastern Region.

The corporation has recently, taken the responsibility of marketing agro-horticultural inputs like fertilizer, pesticides, seeds, agricultural tools and equipment, keeping in view the overall agro-horticultural development of the region.

In addition to providing marketing assistance to small and cottage scale fruit processing units of North Eastern Region, NERAMAC has commissioned a fruit juice concentration plant in 1988 to market low volume high valued products of the region. It is also making attempt to generate employment by providing juice vending machines on franchise basis. This scheme is aimed at, apart from employment generation, creation of alternate marketing structure for natural juice and other processed and fresh fruits and vegetables products.

7. Indian Council of Agricultural Research (ICAR) for North Eastern Hill Region

ICAR Research Complex for North Eastern Hill Region was established in 1975 by the Indian Council of Agricultural Research to create an adequate research base for supporting agricultural development in the North Eastern Hilly Region of the country. It is the first institute of its kind set up by ICAR, which encompasses all the discipline of agriculture, horticulture, animal sciences, agricultural engineering, agroforestry and fishery to cater to the research needs of the tribal areas of N.E. Hilly region including Sikkim. A trainers training centre has also been established to cater the training needs of the entire region. Considering the entire N.E. Hilly region as one unit, the research centres have been so located as to represent the varying altitudes (60m - 1800m above MSL) and agro-climates of the region. The research findings of the institute, at different centres, can thus be utilised for specific altitudinal range and agro-climatic conditions in the component states. The headquarters at Barapani near Shillong has been established in spacious new buildings with well equipped laboratories and administrative wing.

ICAR has the research thrust to evolve suitable integrated farming systems for the hills of the region to replace jhuming (Shifting Cultivation) for increased productivity and for the improvement of citrus plantations to rehabilitate citrus industry.

8. North Eastern Development Finance Corporation Ltd. (NEDFi)

The North Eastern Development Finance corporation Ltd. (NEDFi) was originally conceived in the budget speech of 1995-96, by the Finance Ministry and was promoted by the country's premier public sector financial institutions and bank and incorporated in August, 1995 with its headquarter at Guwahati. The corporation has an authorised capital of Rs. 500 crore and an initial paid up capital of Rs. 100 crore. Further resources will be raised as and when needed. NEDFi started functioning since March 1996. NEDFi is a public financial institution under section 4-A of the Companies Act, 1956.

NEDFi finance a wide spectrum of activities. It finances industries including agro-processing, mining, shipping, transport, tourism, information technology, medical and health services, leasing, generation and distribution of electricity, setting up and development of industrial estates and other commercially viable infrastructure facilities. Agricultural, horticultural plantation will also be financed by NEDFi. Special emphasis is being given to the Information Technology (IT) sector.

9. District Industries Centres (DICs)

Administration of industrial development programmes have constantly been under review to hasten the process of industrialisation and to minimise the hurdles faced by existing and prospective entrepreneurs. Over years, experience has been to bring the network of institutions operating in the field as close to entrepreneurs as possible. Some of the deficiencies experienced in the earlier years prompted the introduction of an innovative scheme known as 'District Industries Centre programme' in the country in pursuance of the Industrial policy statement of 1977.

The DIC is a dynamic institution at the district level which provides all the services and facilities to the entrepreneurs and artisans right from the selection of items for manufacture, supply inputs like credit, raw materials, power, land and building etc. to set up small and village industries in the districts.

Some of the important roles played by the DICs for rapid industrialisation in the district are :

a) to survey existing, traditional industries, raw materials, and human resources in the district.

b) to identify potential industries, identify products and give market forecast for different items and to prepare sample techno-economic feasibility reports.

c) to assess machinery and equipment requirements for various types of identified small-scale and village industries, locate sources for availability for the same and advice the entrepreneurs accordingly.

d) to organise marketing outlets and market survey.

PART - III

Design of the Study

The hills and valleys of North East India can grow different kinds of fruit and vegetable crops. Fruit squashes particularly of pineapple, orange and lemon have a great demand for internal market as well as external market. These fruit products are imported from outside the region due to less production in these states. Even from Myanmar fruit products are imported to North Eastern Region and a variety of fruit products from Myanmar is available in almost all the markets of this region. Canned products of pineapple, on account of its taste and quality, has a great demand from outside states. Fruit processing units of the North Eastern Region still could not meet the demand within and outside the region which is a great loss to the state revenue. Time and again, the feasibility study as well as the detailed project reports have shown this industry is viable in the states of this region in the wake of horticultural development in the states. But till today the question - "Why is the fruit processing industries still unable to fulfil the dream of the people of the region?", remain unanswered. Researcher, therefore, has conducted this study, taking into confidence that by evaluating the past and present performance of the various processing units of the region, strategy and plan of action for greater achievement and for development of processing industry in the region may be made possible. The outcome of this study will boost not only the state and regional economy but also the economy of the country.

Literature Review

Reviewing literature on fruit processing sector was found to be a tough ordeal. Being at the lowest segment of the food processing industries, the sector has not received adequate attention, particularly in North Eastern Region. A brief review of past researches relevant to the present study has been incorporated as follows :

Vipla Chopra (2002)[20] stated that increased utilisation of fruits and vegetables to make different types of processed products is an important way to stabilize process, increase product availability during off season, reduce wastage and utilize the fruits and vegetables as an instruments of industrialisation. The author also stated that there is an acute need for harmonising the existing food laws and to bring about a development orientation to facilitate faster growth of the industry. Girdhari Lal, G. S. et. al (1998)[21] has worked on preservation of fruits and vegetables which is purely technical oriented. The authors highlighted the different methods of preserving fruits and vegetables in all the levels of preservation.

Tata Consultancy Services (TCS) Kolkata and North Eastern Industrial and Technical Consultancy Organisation Ltd. (NEITCO) Guwahati (1998)[22] has jointly studied the export potential of horticulture and floriculture in North Eastern Region. The main observation was that the horticulture sector has not yet received due importance in the North Eastern economy that is largely dependent on subsistence farming. Further, it was observed that only a meagre 3.6 per cent of cultivable land in the North Eastern States is under horticulture production. Maini and Anand (1996)[23] has stated that because of high productivity and value, fruit and vegetable provide much better economic returns per unit area compared to cereal crops and hence it is also a good source of foreign exchange earnings. The authors suggested that a new orientation is needed to be given to the fruit processing industry to upgrade nutrition, minimise post harvest losses, ensure remunerative returns to the growers, increase employment avenues in the countryside and generate foreign exchange earnings.

Singh, H.P. (1995)[24], stated that there is a strong need to develop horizontally and vertically integrated farmers' organizations to undertake various post-harvest operations of fruits and vegetables with professional approach on economic basis and at a commercial scale in order to reduce the existing high marketing cost, enhance the producers' share in consumers' rupee and supply the fresh fruits and vegetables and their processed products to the consumers at a comparatively low price. He also stated that it has now become more conducive to modernize the existing fruit and vegetable processing technology in order to produce more competitive products for export. Bhowmick (1994) [25] has dealt with the marketing of fruits and vegetables in North Eastern Region. He has stated that marketed surplus of fruits and vegetables is the actual amount of produce sold out of the year's production irrespective of his requirements for family consumption, wastage and other payments.

Ex-Fertilizer Servicemen Marketing & Consultancy (P) Ltd. (1994) [26] stated that marketing channel should be widened, intensified and improved in Manipur. The team recommended the formation of a Horticulture Development Corporation which is yet to be implemented in the state. The team also has reported that 30-50 per cent of the horticultural produce in the state were wasted due to transport bottleneck and lack of storage facilities. Th. Biren Singh (1993)[27] has worked on agricultural and horticultural development in Manipur during the plan period but not particularly on fruit processing development.

A feasibility report was prepared by NEITCO Guwahati[28] sponsored by N.E.C. Shillong in 1991. The report contain the Techno-economic feasibility report for fruit processing unit to be set up in Darrang district of Assam. The report stated that there is a feasibility for setting up of fruit processing unit with a capacity to process 3 tonnes of fruit per day. Nurun Nesha (1989)[29] has worked on the development of horticulture in Manipur during the plan period. The research area covers the overall horticultural development in Manipur but does not deal particularly with the fruit processing industry in the state. N.E.C. Shillong has also sponsored a feasibility report prepared by NEITCO Guwahati (1988)[30]. The report was discussing for revival of fruit processing unit at Vairengte, Mizoram which stated that after revival the plant will break even at 62 per cent of the capacity and earn profits from the second year of its operation.

Ghosh (1985)[31] has dealt with horticulture in North Eastern India where he has discussed horticulture industry in general but no specific work has been done on fruit processing industry by him. Centre for Food Technological Research Institute (CFTRI), Mysore (1978)[32] has conducted a study on packaging and transportation methods for the horticultural products of the North Eastern Region. The study concluded that there is a scope for improving the quality of the fruits transported to Kolkata and other prospective market.

Though the present review may not be the exhaustive review of literature, but it can be concluded from the above review that there was no in-depth study so far conducted in this field particularly in the North Eastern States. Hence, the present case study of fruit processing units will have a greater significant for the development of fruit processing industry particularly in the North Eastern Region.

Objectives of the Study

The main objectives of the present study is to critically examine and evaluate the present status, constraints, and future strategies/plan of action, keeping in view the production and marketing potential, for the development of fruit processing industry of the North Eastern Region in general and particularly Manipur and Assam.

More specifically, the study will have, iner-alia, the following objectives :

1. To evaluate the performance of all the existing fruit processing units in Manipur and Assam and to identify their problems and prospects.
2. To critically examine the production and marketing strategies of different units in Manipur and Assam.
3. To analyse the role of state and central government toward the development of fruit processing units in the region.

4. To offer constructive suggestions and guidelines to remove the problems associated with the processing industries in Manipur and Assam for further development and improvement.

5. To identify the viability and suitability of fruit processing unit in the hill districts of Manipur.

Hypothesis : The present study is based on the following hypothesis:

1. Production per unit cost of processed fruits is high due to high cost of transportation involved in purchasing the fruits.

2. Most of the fruit processing units are born sick due to lack of basic infrastructural facilities such as power, transport, communication and banking facilities.

3. More than 50 per cent of the total market is captured by the product which is manufactured from outside the state/region.

Selection of the Units

The present study confines itself to the list of fruit processing units obtained from the office of the Deputy Director (F&VP) Eastern Region, Kolkata. Out of the total 61 units given, all the units except units producing sweetened aerated water were taken for study.

Hence, out of the 13 units in Manipur, 9 units only were selected for interview as two of the units are dealing with sweetened aerated water and packaging, one unit is just the combination of all the private units in the state which is known as a relaballer, is insignificant at present. The remaining one unit could not be contacted. Hence, only 9 units were interviewed. And out of this 9 units, one unit is found to be sick. From Assam, out of the 25 units, 21 units which were processing fruits were selected for interview but only 8 units could respond to the said interview. Again out of this 8 units, 4 units are sick, 1 unit is already converted into bakery business. So, only 3 units performance could be evaluated.

Along with the units of Manipur and Assam, the researcher has established a contact with the units of the other North Eastern States through despatching questionnaire to them. Out of the total 8 units in Meghalaya, 7 units were selected but only 2 units responded with the filled questionnaire despatched to them. Mizoram having 2 units both were selected and 1 unit promptly responded. All the 3 units of Arunachal Pradesh and 4 units of Tripura were selected for interview but none of the unit from this two states responded to the dispatched questionnaire. Finally, out of the 6 units in Nagaland, 5 units were selected for interview but none of the unit responded.

Thus, the present study is based on 14 fruit processing industry of the region i.e. 8 units from Manipur, 3 units from Assam, 2 units from Meghalaya and 1 unit from Mizoram.

Methodology

The present study is based on both primary as well as secondary data. Primary data have been obtained with the help of personal interviews with the concerned executive of the units and also with the help of interview questionaire sending to the selected units.

Secondary data for the study have been collected from the annual reports of the Ministry of Food Processing Industries, Government of India and the annual balance sheet of the units. The other data have also been gathered from various journals, reports, newspapers, periodicals and unpublished work of the scholars in this field.

With the help of the data contained in financial statements, various selected statements were prepared and calculated with various statistical tools and techniques such as percentage analysis, ratio analysis, composite growth rate, correlation analysis and significant test through T-test and Chi Square test. In order to give the conclusions and findings, a comparison have been made. Apart from this with the help of the conclusion drawn, suggestion have been made.

Interview Questionaire

An interview questionnaire having 61 questions have been framed to ensure that the investigator gets the desired information. This questionaire consists of five blocks. The first block covers general information such as name and address of the sample unit, Registration No., FPO Licence No., date and year of commencement, brand name and the objectives of the processing unit. The second block consists of the detailed production information such as types of machine, capacity of machine, types of fruits and vegetables used, suppliers, prices of this raw materials, purchasing factor, other ingredients used, persons employed, working days and working hours, quantity of fruit consumed, types of products produce, quality control, by-product, and problems of production. The third block covers marketing outlet, time lag for collecting amount of credit, distribution channel, methods of pricing, unsold stock, sales promotion, packaging, mode of transportation, area of competition, trade fairs and problems in marketing.

The fourth block relates to the financial aspects such as, fixed and working capital, cost of factory buildings, time lag for converting raw material into cash, financial help, depreciation policy, problems of getting finance and repayment. The five block includes some general problems such as implication of industrial policy, power supply, transport facilities, banking facilities, training facilities, choice of location and product competitiveness.

Reference period

The present study has been selected for a period of ten years commencing from 1991 to 2000. This period is quite satisfactory to evaluate the performance of all the existing fruit processing units of the region.

Limitations of the study

The present study has various limitations. No sufficient literature are available specially the data pertaining to the North Eastern Region as sufficient study in this field not have been conducted so far. So, the study has to rely more on the primary data to which the questionnaire were being prepared. Unfortunately, only few units from Assam, Meghalaya and Mizoram could respond for the purpose after much pursuasion and personal request. None of the units from Arunachal Pradesh, Tripura and Nagaland responded to the questionnaire inspite of fervent request made by the investigator. It is from the state of Manipur, where the investigator belong, that maximum response was received. Some units were very old that they could not traced their history of beginning. Hence, the analysis is done on the basis of the available limited informations and data.

Significance of the study

Though the present study has various limitations, the outcome of the study will be a good reference for future studies. The outcome of the study may be highly useful to the investors, creditors, banks and other financial institutions to take various investment decisions in this sub sector of food processing industries. It will also provide appropriate strategies to the concerned management to ensure the fruitful results i.e. production, marketing and financial performance of the units. It will also help the state and the central government in allocating fund for horticultural development in general and fruit processing industry in particular.

Processing of fruits will increase the earning of the farming sector. This industry is a labour intensive one which provides employment to 19 per cent of the industrial workers. It accounts for about 19 per cent of the gross domestic product, 7 per cent of the total industrial investment. Fruit products are export-oriented and hence it can earn more foreign exchange through this industry. The outcome of this study will boost not only the state and regional economy but also the economy of the country.

Chapter Planning

Chapter - one includes three parts. The first part deals with the introduction of food processing industries in India. It also includes fruit crops in India and the profile of the North East. The second part deals with the role of the government towards the development of fruit processing industries where the national policy on food processing and the industrial policy for the North Eastern Region for the development of this sector are being highlighted. The third part includes objectives of the study, research design and hypothesis on which the study is based and the limitation and the significance of the study.

Chapter - two is divided into two parts. The first part deals with the present status of fruit processing industries in India in general and the fruit processing industry in the North Eastern Region in particular. The second part covers the infrastructural facilities such as transport, communication, power and banking and finance available to the North Eastern States are highlighted which is one of the important area where hypothesis of the study is based.

Chapter - three is grouped into three parts again. The first part discuss the production and marketing of fruits in India along with the export trend of processed fruits. The second part deals with production and marketing of processed fruits in North Eastern States. The third part highlights the viability and suitability of horticultural industry in Manipur.

Chapter - four is divided into five parts. The first part includes the historical background of selected 14 fruit processing units of the region. The second part deals with the data analysis of the selected units with the help of various statistical tools. The third part covers the financial analysis with the help of ratio and ranking analysis. The fourth part includes the SWOT analysis of fruit processing units which discusses the strengths, weaknesses, opportunities and threats of fruit processing industry. Finally, it also covers the hypothesis testing of the present study.

Chapter - five includes two parts. The first part covers the findings of the study. The second part includes the suggestive measures to solve the prevailing problems.

REFERENCES

1. Prusty, S.R., Food Processing Industry, *SBI Monthly Review*, February 2000. p. 83
2. *Ibid*, pp. 83-84.
3. Shete, N.B., Problems and Prospects of Financing Agro-processing Industry in India, *SBI Monthly Review*, February 2000. pp. 57-58.
4. *Ibid*, p. 58.
5. Prusty, *op. cit.*, p. 84.
6. Secretary, Department of Food Processing Industries, Govt. of India.
7. Prusty, *op. cit.*, p. 85.
8. Prusty, *op. cit.*, p. 85.
9. Manager, Economic Research Department, SBI, Corporate Central, Mumbai.
10. Prusty, *op. cit.*, p. 86.
11. Prusty, *op. cit.*, p. 86
12. Prusty, *op. cit.*, pp. 86-87.
13. Sudheer and Dash, Some Strategies to Boost Fruits and Vegetables Sector, *Yojana*, November 1988, p. 37.
14. *Ibid*, p. 37.
15. Singh, H.P., Improving Post Harvest Technology of Fruits and Vegetables, *Yojana*, December, 1998. p. 35.
16. Deputy Agricultural Marketing Advisor, Directorate of Marketing and Inspection, Nagpur.
17. Dhirendra Nath Borthakur, *Agriculture of the North Eastern Region*, Deecee Prakashan, Guwahati, 1992, p. 1.

18. *Ibid*.
19. Department of Agriculture & Cooperation, Government of India.
20. Vipla Chopra and Food Processing Industry in India, *Political and Economic Journal of India*, Vol. 11. No.1, Chandigarh, 2002.
21. Girdhari Lal, G.S. et.al, Preservation of Fruits and Vegetables, Publications and Information Division, ICAR New Delhi, 1998.
22. APEDA, Guwahati : Export Potential Assessment of Horticulture and Floriculture Products for the North Eastern States : A Survey Conducted on behalf of APEDA, Guwahati, 1998.
23. S.B. Maini and J.C. Anand, Fruit and Vegetable Processing Industry. Present status and future prospects, *Productivity*, Vol. 36. No. 4, Jan.-March 1996.
24. Singh, H.P., Fruit and Vegetable Processing Industry in India, *Yojana*, Vol. 38. No. 24, January 15, 1995.
25. B.C. Bhowmick, Marketing of Fruit and Vegetables in North Eastern Region, Paper presented in the Training Organised by NIAM, Jaipur on Agricultural Marketing Practices in North Eastern Region (Dec. 6-9, 1994) at Shillong.
26. National Horticulture Board, Techno-Economic Feasibility Study for the Development of Horticulture in the State of Manipur : A Study Conducted by Ex-Fertilizer Servicemen Marketing & Consultancy (P) Ltd. Kolkata and was Sponsored by National Horticulture Board, Ministr y of Agriculture, Govt. of India.
27. Th. Biren Singh (Ph.D Thesis), A Perspective of Agricultural and Horticultural Development in Manipur During the Plan Period, M.U. 1993.
28. N.E.C. Shillong, Techno-economic Feasibility Report for a Fruit Processing Unit in Assam. Report prepared by NEITCO, Guwahati, 1991.
29. Nurun Nesha, The Development of Horticulture in Manipur During the Plan Period, (M.Phil Dissertation), M.U. 1989.
30. N.E.C. Shillong, Feasbility Report for Revival of Fruit Processing Unit at Vairengte (Mizoram). Report prepared by NEITCO, Guwahati, 1988.

31. Ghosh, S.P., *Horticulture in North Eastern India*, Associated Publishing Company, New Delhi, 1985.

32. N.E.C. Shillong, A Study on Packaging and Transportation Methods for the Horticultural Products of the North Eastern Region. A Study Conducted by CFTRI, Mysore, 1978.

2

Present Status of Fruit Processing Units

Introduction

Fruit processing industry is considered to be the backbone of horticulture industries as it take care of gluts and all the wastes occurring, inspite of all improvements in the distribution and marketing of fresh produce. The agragrian base of India holds great promise with regard to processing of fruits and this segment of food processing industries is expected to grow fairly at high speed in coming years. The fruits because of their high productivity and value offer much higher economic returns per hectare acreage compared to cereal crops. They also have good export potential.

The prominent items processed in the fruit sector are fruit pulps and juices, fruit-based ready to serve beverages, canned fruit jams, squashes, ketchups, and pickles. More recently the industries has taken up the processing of freeze-dried fruits, fruit concentrates, and canned pineapple slices. This canvas of processed fruits is likely to expand in coming years and include minimally processed fruit pieces and slices, hurdle technology (HT) stabilized fruits besides other ready to eat more cost-effective substitutes for canned products. The progress of fruit processing industry in India is rather slow comparing with other countries.

Progress of the Fruit Processing Industry in India

It was in the first World War that some mechanization and commercialisation entered into this industry mainly to meet the demand of the armed forces. The second World War gave it a much-needed fillip. Since then it has been progressing fairly well and has been meeting the entire local demand and in a very limited way entered the exports market. It is, however, not a heavy weight industry but has the potential to develop into a 'Sunshine Industry' of the country. Inspite of the fact that India, is the second largest producer of fruit in the world yet the commercial processing of fruits is less than 2 per cent. The main reason behind is that domestic consumption of processed items is quite meagre because of economic reasons and also as a matter of habit. The Indian consumers by and large very much prefer fresh fruits and vegetables. The high cost of packaging pushes up the cost of the processed items and thereby makes them out of reach of the common men. Because of the varied agro-climatic conditions some fresh fruits are available throughout the year. The fruits like bananas are non seasonal and apples, oranges, and potatoes etc. are put in the cold stores thus prolonging their shelf life and making them available in the off season too. Some fruits like guavas, oranges have two seasons so they are available in fresh form for four to five months in a year.

Processing Units and Installed Capacity

Presently there is a little over 5293 units registered under the Fruit Products Order (FPO) of 1955 distributed all over the country. Most of the units fall in the cottage and small-scale sector. A few modern processing plants have now come up and many more are in the pipeline. The installed capacity which was 11.08 lakh tonnes, in December 1993 increased to 21 lakh tonnes at the end of the year 2000.[1]

Modern Units

After the liberalisation of the economic policies in the

country in 1991, a few modern plants to produce mango pulp, tomato paste etc. in aseptic packing, freeze drying of many fruits is being taken up. It is expected that in the years to come many modern state of the art plants shall come up, in food processing industries particularly in fruit industry.

Joint Ventures

Since liberalisation in July, 1991 till February, 2000, 1120 proposals of industrial licenses and 100 per cent export oriented units were approved and about 248 such proposals have already been implemented[2]. The important countries with which the Joint Ventures have been signed are U.S.A., U.K., Netherlands, Switzerland, and Germany. The proposals include the fields like technology transfer, financial and marketing tie-ups. These tie-ups include production of items like banana and mango puree, fruit concentrates. A few proposals of frozen fruits and vegetables have also been approved.

Product Range

The important items manufactured in the country are fruit pulps particularly of mangoes, ready to serve juices, canned fruits, jam, pickles, squashes etc. Recently, items like frozen fruits, pulps, dehydrated and freeze dried vegetables, canned mushrooms etc. are also being produced. In the coming years new items like carbonated fruit drinks, dehydrated and freeze dried fruits, fruit juice concentrate are expected to be manufactured.

Exports

India in a small way has been in the export market for almost 30 years. Among the popular items in export are mango chutneys, pickles, fruit juices, frozen and canned fruits . In the year 1997-98 the exports of processed fruits and vegetables were in the order of 299 thousand tonnes valued at Rs. 761 crore or US $ 200 million [3].

The main markets for mango pulp are Saudi Arab, Kuwait, UAE, Netherlands & Hongkong. In case of pickles & chutneys the popular markets are USA, UK, UAE, Germany, and Saudi Arab. Other items like tomato paste, jams, jellies and juices are exported to USA, Russia, UK, UAE and Netherlands.

Quality Control (ISO 9000)

There is an all round realization at every level that quality production of international standard is the watchword for survival now. Consequently, the processor, small or big, have started taking steps to improve quality right from the selection of the raw material and that of the ingredients. Side by side steps have also been taken by them to upgrade the hygienic and sanitary conditions of the workers, plant and machinery so as to ensure quality of the finished product. There is a much greater emphasis now to invest in research and development and also in product innovation. The Agricultural and Processed Food Products Export Development Authority (APEDA) gives subsidy to upgrade the laboratory facilities etc. It has also brought out a book entitled "ISO 9000 on The Food Industry : A Practical Guide". It deals with the entire spectrum of quality control, e.g. documentation, installation, certification of quality management system etc.

Fruit Processing Industry in North Eastern Region

Characterized by the prevalence of tropical, sub-tropical and temperate climate and supplemented with high rainfall and favourable soil conditions, the North Eastern Region is highly conducive for the cultivation of diverse and exotic horticultural products. The fruits crops of the region vary from tropical fruits like litchi, papaya, orange, lemon, pineapple and banana to apples, plum, peach, pear and apricot predominantly found in the temperate zones. In fact, there is hardly any fruit which cannot be grown in North Eastern Region of India.

According to APEDA'S "Export potential assessment of horticulture and floriculture products for the North Eastern

Table - 2.1
Fruit crops in the North Eastern Region (1996-97)

Area - thousand hectares
Production - thousand tonnes

Fruit crops		*Assam*	*Arunachal Pradesh*	*Manipur*	*Meghalaya*	*Mizoram*	*Nagaland*	*Tripura*	*Total of North East*
1		*2*	*3*	*4*	*5*	*6*	*7*	*8*	*9*
Pineapple	A	13.99	6.63	9.95	8.79	1.03	1.81	5.02	47.22
	P	207.84	29.92	69.65	77.2	7.15	44.17	44.47	480.4
Mandarin	A	5.9	5.57	0.83	—	7.02	1.27	4.48	25.07
Orange	P	68.53	11.84	3.32	—	28.07	18.03	21	149.79
Other citrus	A	7.37	2.72	0.97	6.96	0.96	1.16	2.71	21.69
Lemon/Lime	P	40.64	3.98	5.89	32.98	2.79	16.55	3.52	91.8
Banana	A	41.34	2.65	1.41	5.37	2.81	1.51	4.93	60.02
	P	595.26	9.5	10.72	63.97	14.87	23.36	33.63	751.31
Mango	A	1.36	0.08	0.16	—	0.48	0.17	5.02	7.27
	P	7.81	0.06	0.88	—	2.56	0.46	41.27	—
Guava	A	3.08	0.68	0.7	—	0.34	0.45	—	4.25
	P	40.82	2.61	2.1	—	1.32	3.23	—	50.08

(Contd...)

1		2	3	4	5	6	7	8	9
Litchi	A	4.06	0.18	—	—	0.08	0.55	5.42	10.29
	P	17.56	0.25	—	—	0.27	0.17	2.79	21.04
Papaya	A	7.48	0.46	1.85	0.48	0.3	0.29	—	10.86
	P	110.4	1.94	10.18	3.93	2.43	2.48	—	131.36
Jackfruit	A	18.4	0.12	0.26	—	0.22	0.23	8.12	17.35
	P	160.2	0.71	1.04	—	1.39	0.91	214.58	378.83
Apple	A	—	2.37	—	—	0.03	0.08	—	2.48
	P	—	5.41	—	—	0.03	0.45	—	5.89
Pears	A	—	0.58	—	—	0.09	0.75	—	1.42
	P	—	1.67	—	—	0.44	2.46	—	4.57
Plum &	A	—	0.73	0.61	—	0.18	0.93	—	2.45
Peach	P	—	1.9	3.97	—	0.52	3.43	—	9.82
Total	A	92.98	22.77	16.74	21.6	14.54	9.7	36.66	209.37
	P	1259.06	69.79	107.75	178.08	52.04	115.7	383.66	2,127.13

Sources : 1. Basic Statistics of North Eastern Region 2000. (NEC Shillong)
2. APEDA Guwahati

Note : A = Area P = Production.

State's Report, 1998"; the horticultural sector has not yet received due importance in an economy that is largely dependent on subsistence farming. Only one per cent of the total geographical land is under cultivation of horticultural produces where the cultivable area is around 26 per cent. The main fruits of the region are banana, citrus fruits (including orange) and pineapple. Banana accounts for 25 per cent of the area under horticulture crops, while citrus and pineapples have a share of 21 per cent and 19 per cent respectively. The other fruits cultivated in this region are jackfruit, apple, litchi, guava, papaya, mango, plum, peach, pear and others. Some rare fruits like passion fruits having good export potential are cultivated in Mizoram, Nagaland and Senapati district of Manipur. Though these fruits are not much in demand in the local market, but they have a good demand in international market. In recent years, the Government of Mizoram, Nagaland and Manipur have realized the potential and are encouraging the cultivation of passion fruit. However, production has to increase by many folds for the fruits to earn an exportable status. Most of the north eastern states like, Tripura and Meghalaya, the fruits grown in the border areas are exported almost in totality, as the cost of selling inland is often more than the prevailing market price. In such places the export market is merely seen as the extension to the local market and the trade carried on is basically unofficial in nature.

The area and the production figures of fruits are given in Table-2.1 above. On a rough estimate, the area under fruit crops in this region is about 209.37 thousand hectares with an estimated production of 2,127.13 thousand tonnes of fruits. Among fruits, production of banana rank first with 751.31 thousand tonnes followed by pineapple with 480.40 thousand tonnes, jackfruits with 378.83 thousand tonnes and oranges with 149.79 thousand tonnes.

The nature of horticulture in most of the hilly states of North - Eastern Region is consumption oriented, traditional horticulture and in some related pockets or areas gradual transition is taking place from subsistence horticulture to market oriented commercial horticulture.

Among the horticultural crops, fruits are highly perishable commodities and due to lack of adequate post harvest handling facilities and proper infrastructure in the region, post harvest losses due to spoilage are very high. Actual post harvest losses of fruits have been estimated to be as high as 30-40 per cent in the region[4].

Hence, processing has a special significance in the region because of the high rate of spoilage on account of lack of storage facilities and inaccessibility to far-flung market centres. But the processing scenario of the entire region is far from satisfaction. As per the list provided by the Department of Food Processing Industries, Eastern Region, Kolkata, till November 1999, there are only 61 units of fruit processing in the region shown in the below Table 2.2.

Table- 2.2
No. of Fruit Processing Units in the North Eastern Region

State	*Public*	*Co-operative*	*Private*	*Total*
Arunachal Pradesh	3	Nil	Nil	3
Assam	2	1	22	25
Manipur	2	2	9	13
Meghalaya	2	Nil	6	8
Mizoram	1	Nil	1	2
Nagaland	2	1	3	6
Tripura	2	1	1	4
Total	**14**	**5**	**42**	**61**

Source : List of NE Region Fruit Processing Units; Deputy Director (F &VP) Eastern Region, Kolkata Ministry of Food Processing Industries, Govt. of India.

As many as 25 units are in Assam, followed by 13 units in Manipur, 8 units in Meghalaya, 6 units in Nagaland, 4 units in

Tripura, 3 units in Arunachal Pradesh and 2 units in Mizoram. The present status of these units are shown in the appendix - 1. More detailed study with the help of the above table-2.2 shows that there are 14 public sector unit which are run and own by the public, 5 units are under co-operative sector and 42 units i.e. 68.85 per cent of the existing units in the region are run under private ownership. When we look at the figure of Assam, Manipur, Meghalaya and Nagaland, indicated in the above Table, it shows that out of 25 units in Assam 22 units i.e. 88 per cent of the units in the state are under private sector, in Manipur out of 13 units 9 units i.e. 69.2 per cent of the total units belongs to private sector, and in Meghalaya also out of 8 unit 6 units i.e. 75 per cent of the total units existing in the state are run by private entrepreneurs. In Nagaland also 60 per cent of the total units are under private sector. Therefore, it is obvious to say that private sector plays an important role in the processing of fruits in the North Eastern Region. On the other hand the role of co-operative sector is negligible as its contribution to this sub-sector is just around 8 per cent in the region. This could be one of the reason for slow progress of processing units in the region where there is vast potentiality in growing horticulture crops particularly the fruit crops in the North Eastern Region.

PART - II

Infrastructural Development in North Eastern Region

Rapid development of the economy depends on adequate investment in infrastructure, a key issue here is imposition of appropriate user charges necessary to provide adequate returns on investment. There is a common belief that the North Eastern Region of India is the most backward region in infrastructure development such as power, transport, and communication. The Ministry of Finance, Government of India appointed a group headed by Dr.(Mrs) I.K. Borthakur to examine the problems of industrialisation of the region in 1994, it has made the following observation "the group felt that the basic infrastructure in the North Eastern Region was not as bad as generally believed. On the contrary the region was far better than most of the other states of the country in terms of per unit population and per unit of geographical area served by the national highways, railways, power and communication. However, most of the facilities were under utilized."[5]

And to quote late G.M.C. Balayogi speaking about the problems of the North East while inaugurating the North East Regional Commonwealth Parliamentary conference at Gangtok on May 22, 2001, he said "The problems of the North East stem from lack of connectivity dearth of basic need, inadequate infrastructure facilities, socio-economic backwardness and insurgency, while all these factors have led to inhibited development of the region, perhaps the most crucial has been the insurgency."[6]

The present discussion confines itself into the four basic area of infrastructure viz. power, transport, communication and banking and finance, endeavours to highlight the present scenario with the comparison among the north eastern states and with that of all India existing position.

Transport

An adequate and efficient transportation network is essential for the economic growth of a region. Road is the major means of transportation in the whole North Eastern Region. The presence of railway network, which is one of the economic means of transportation in the region, is negligible except in Assam which has 2373.52 Km. railway network. However, in the matter of road length, North - Eastern states like Tripura, Assam and Nagaland are above all India average. The total road length per 100sq. km. of Tripura is 140.4 per cent of Assam is 86.8 per cent and that of Nagaland is 82.8 per cent, against 73 per cent for the country as a whole but the percentage of the whole North Eastern Region is just 52.08 which is less than the all India average as the percentage of the other state in the region are comparatively low, as it is shown in the Table - 2.3 below.

Further, longevity and quality of road of the region as compared to other states of the country is very poor.

Communication

The development of industries without better communication system becomes impossible for continuity . The communication through postal, couriers, telecommunication are vital for the development of industries. The present status of communication in the region compared to national average is poor. Population per post office of the region is 3555 against 5502 of all India average as indicated in the below given table. Though the population per post office is less, about 75 per cent of the region's geographical area are hilly and it is difficult for prompt delivery and easy communication. Therefore, the number of post offices should be increased to reduce in population per post office with better telegraphic and telecommunication facilities. Population per Telecommunication exchange in North Eastern India is 53391. The ratio indicated for Arunachal Pradesh (13471) and Mizoram (19301) are satisfactory comparing with Manipur and Assam having 72383 and 64189 persons respectively per telephone exchange. More telephone exchange should be installed to reduce number of persons per telephone exchange. The private communication network by way of courier, STD/ISD/E mail/Internet and other information technology should be encouraged to reach the people with least effort.

Power

Power is the backbone of the industries and plays vital role in fostering industrial activity. The per capita consumption of power during 1995-96 in the region compared to national average is very poor.[7] The above Table - 2.3 shows that Meghalaya state has the highest per capita consumption of power in the whole region with 143 KW against the all India figure of 335.42 KW. But the state like Tripura and Arunachal Pradesh has the lesser per capita power consumption of 73KW

Table - 2.3
Selected Infrastructural Indicators of North East India

	Parameters	Arunachal Pradesh	Assam	Manipur	Meghalaya	Mizoram	Nagaland	Tripura	NE.	All India
1	Transport :									
	- Road per Sq. Km.	12.2	86.8	48.2	37.4	32.8	82.8	140.4	52.08	73
	- Railway Network (Km)	1.26	2373.52	1.35	—	1.5	12.85	44.72	2435.2	—
	- Road per 1000 population	11.38	3.05	5.98	4.66	9.87	11.44	5.45	4.23	2.85
2	Communication									
	- Population per post office as on 31-03-99	3,028	5,814	2,719	3,682	1,783	3,974	3,865	3,555	5,502
	- Area served by a post office (Sq. km.)	294.72	22.45	33.14	46.86	54.52	53.74	14.66	—	21.42
	- Population per Telecommunication Exchange as on 31-10-99	13,471	64,189	72,383	42,705	19,801	52,333	58,021	53,931	—

(Contd...)

3	Power									
	- Installed (MW) (Hydro)	23.6	2	2.8	185.2	5.4	3.3	16	—	—
	- Potential (MW) (Hydro)	26,756	351	1,070	1,040	—	1,455	176	—	31,857
	- Per capita power consumption (1995 - 96) KWH	78	100	118	143	128	79	73	—	335.42
4	Banking and Finance (as on 31.03.99)									
	- No. of Scheduled commercial banks	198	1307	104	216	—	95	223	—	—
	- Population (in 1000) per bank	17	21	28	13	12	22	20	—	15
	- Aggregate deposit (Rs. In Lacs)	55,874	73,472.55	43,674	1,35,859	35,025	68,457	1,10,245	11,84,607	—
	- Aggregate Advance (Rs. In Lacs)	10,754	2,39,849.9	20,787	25,964	8,195	14,547	38,786	3,58,882.9	—
	CD. Ratio	19.25	32.61	48	19.11	23.4	21.25	35	30.29	55.5

Sources : 1. Basic statistics of North Eastern Region, 2000. (NEC Shillong)
2. North Eastern Institute of Bank management, Guwahati, Assam.
3. NE Newsletter vol. 1 No. 6 May 2000, Ministry of Home Affairs.
4. North Eastern at a glance, NEC, Shillong.

and 78 KW respectively, which is the very indication of backwardness in the industrial development of the region. Out of the seven states in the region, Meghalaya is the only state which could meet their power supply requirement having an installed hydro power capacity of 185.20 MW where as, the remaining states purchases power from other states to fill the requirement and very often power cut is taking place in many of these states. Arunachal Pradesh has great potential for power resources but due to lack of initiative taken by the state government, NEC and the central government, the potential hydro power resources of Arunachal Pradesh are lying as sleeping beauty.

Banking and Finance

Needless to mention, finance is the main pre-requisites of every productive operation. The very success or otherwise of an industrial unit depends considerably upon this factor, for many of the business problems find their solution on financial strength of the unit.

Adequate and timely supply of finance at the least cost of capital are the most essential factors to minimise gestation period, avoid project cost escalation and their fruitful co-ordination for industrial growth and development. Finance for the industrial development schemes can come only from a network of banking facilities. There were 1307 banks in Assam, as on March 1999, followed by Tripura (223) and Meghalaya (216). The remaining states like Arunachal Pradesh , Manipur and Nagaland have bank branches of 198, 104 and 95 respectively. Figure for Mizoram is not available. However, the average number of persons served by a bank branch for Manipur is 28,000 against all India average of 15,000 persons per branch. The figure for Arunachal Pradesh, Assam, Meghalaya, Mizoram, Nagaland and Tripura were 17, 21, 13, 12, 22 and 20 thousands respectively.

To conclude, the available infrastructural facilities in the region are inadequate to support the growth of fruit processing units in the region. The state Meghalaya alone has sufficient power to meet the industries developmental requirements, but the transport and communication facilities are still lacking behind as indicated in the above table. Hence, the infrastructural facilities are to be considered and developed first, if the fruit processing units in the region are to be fully developed.

REFERENCES

1. *Annual Report, 2000-01*, Ministry of Food Processing Industries, Government of India.
2. *Ibid.*
3. *Ibid.*
4. Singh, R.K.P., et al, Post Harvest Management of Fruits and Vegetables in N.E. Region. Paper presented during the National Seminar on Management Strategies for North Eastern Hilly Ecosystem. Aug. 1998, Shillong.
5. NERC, IDBI, Guwahati : Proceeding of the meeting held on December 18, 1995 at Guwahati.
6. Press Trust of India (PTI), quoted in *The Sangai Express*, May 22,2001.
7. Rana Bijay Dev : Present Scenario, Problems and Prospects Infrastructure Development in North Eastern Region, Infrastructure Development for Industries, Dibrugarh University.

3

Production and Marketing of Processed Fruit Products

Introduction

In the previous chapter, we have discussed the present status of fruit processing industry in India and the infrastructural facilities available for the development of this sub-sector. We shall now discuss in brief the production and marketing of fruits in India along with its present trend of processed fruits in India and in the North Eastern Region.

According to Food and Agricultural Organisation (FAO) production year book, 1994, India is the second largest producer of fruits after China and its share in World production of fruits is about 9 per cent . It produces 54.20 per cent of World's mango and 15.02 per cent of World's banana. It produced 24 million tonnes of fruits in 1984 - 85 which rose up to 22,808 million tonnes[1] of fruits in 1993-94. The area covered under fruits plantation in 1993-94 was 3.3 million hectare. The world situation and the percentage of total production of major fruits in India are given in Tables-3.1 and 3.2.

It is found out from the Table - 3.1, that mosambi has the highest percentage in the world production of fruits. Next to mosambi is grapes with 21.95 per cent and though India's contribution to the world production of grapes was just 1.33 it stood the first position in the world production of grapes.

Banana became the third highest in terms of percentage of world production of fruits. India's contribution to the world production of banana was 15.02 per cent and stood the 8th rank in the world production of banana. Again, India is the largest producer of mango in the world contributing 54.20 per cent of the world production of 18,450 million tonnes of mango in 1993-94. Finally, India's contribution to the world production of pineapple is 6.93 per cent and stood the 5th position in pineapple production.

Table 3.1
World Situation and India's Contribution to the World Production of Fruits During the year 1993-94

(Production in million tonnes)

Commodity Fruits	*World*	*Percentage to the total production*	*India*	*India's position in the world production*	*India's contribution to the world production in percentage)*
Apple	48,890	19.03	1,238	—	2.53
Banana	52,584	20.47	7,900	8th	15.02
Grapes	56,392	21.95	750	1st	1.33
Mosambi	58,731	22.87	2,100	—	3.58
Mandarin	9,991	3.89	—	—	—
Mango	18,450	7.18	10,000	1st	54.2
Pineapple	11,832	4.61	820	5th	6.93
Total	256,870	100	22,808	—	8.88

Source : Productivity, vol. 36, No. 4 January - March, 1996, p. 557.

As we have seen the world's situation and India's contribution to the world production, of fruits, it is also important to know the percentage of each of the items to the total production of major fruits in India. The Table -3.2 given below shows that out of the total fruits production, banana occupies 32 per cent and accounts for 105.26 lakh tonnes, mango 28 per cent which accounts for 92.10 lakh tonnes, citrus 7 per cent, apple 4 per cent, pineapple 3 per cent, grapes 3 per cent, guava 1 per cent and others accounts for 23 per cent but it is unspecified and so it becomes insignificant in this analysis. Banana and mango occupies the highest rank in terms of percentage of fruit production in India and accounts for 105.26 lakh tonnes and 92.10 lakh tonnes respectively.

Table 3.2
Percentage of Total Production of Major Fruits in India (1992 - 93)

Fruits	*Quantity (in lakh tonnes)*	*Percentage*
Mangoes	92.10	28
Banana	105.26	32
Grapes	6.58	2
Pineapple	9.87	3
Apple	13.16	4
Guava	3.29	1
Citrus	23.02	7
Others	75.65	23
Total	328.93	100

Source : Productivity, Vol. 36, No. 4 January - March, 1996, p. 558.

Fruits are highly perishable commodities and due to lack of adequate post harvest handling facilities such as plucking and storing and proper infrastructure in India, post harvest losses due to spoilage are very high. Actual post harvest have been estimated to be as high as 25-30 per cent of the value of the produce. At this scale, the post harvest losses are Rs. 6000 to Rs. 7000 crore per annum. Even if 10 per cent of this could be saved by converting them into processed products at peak production season, there will be a saving of only Rs. 650 crore to the horticultural wealth of the country. But the fruits processing industry at present is utilizing only about 2 per cent of total production for processing, as against the performance of other countries - Malaysia 83 per cent, Philippines 78 per cent, Brazil and USA 70 per cent each respectively[2].

The estimated production of fruits for 2002-03 in the country is 33 million tonnes or 8 per cent of the total world production next to China[3]. Calculating 30 per cent post harvest losses on the above estimate, the total losses of India will be 9.9 million tonnes. Whereas processing of 2 per cent of fruits in the above estimate will be 0.66 million tonnes. Therefore, the remaining 22.44 million tonnes of fruits are supposed to have been marketed and consumed in fresh form.

It is now obvious that the total losses of fruits in the country is comparatively very high and the processed percentage of fruits are very low. Fruits are highly perishable commodities and so in order to avoid post harvest losses and to boost the utilisation of these food items, there is a need to increase processing of fruits particularly in the country like India. It will not only prevent heavy losses but it will also generate higher income for the country and open larger avenues for employment to unemployed youths.

Present Production Trend of Fruits Processing Industry

We have already discussed above concerning the production and utilisation of fruits in India. It is pertinent now to highlight the present trend of processed fruits in India. The Table-3.3 given below shows, the Fruit Product Order (FPO) Licenses issued, installed capacity, production trend of fruits and growth in production over the previous year for the last ten years in this sub sector as on Ist January every year.

Table 3.3
FPO Licences, Installed Capacity, Production and Growth in Production of Processed Fruit Products in India

Year	*FPO Licences*	*Installed capacity (Lakh tonnes)*	*Production (Lakh tonnes)*	*Growth in production (percentage)*
1991	3925	8.94	2.80	14.29
1992	4057	9.50	3.60	28.57
1993	4132	11.08	4.69	30.28
1994	4270	12.60	5.59	20.00
1995	4363	14.02	6.76	20.93
1996	4674	17.60	8.50	25.74
1997	4932	19.10	9.60	12.94
1998	5112	20.40	9.10	5.20
1999	5198	20.80	9.40	3.30
2000	5293	21.00	9.80	4.26
2001	NA	21.10	9.90	1.00

Source : Annual Report 2001-02, Ministry of Food Processing Industry, Government of India.

NA = Not available.

As can be seen from the above table, there is an increasing trend of number of processing units, installed capacity, production and growth in production of processed fruit products.

The installed capacity of the industry increases from 8.94 lakh tonnes during 1991 to 21.10 lakh tonnes in 2001. The number of licences issued under fruit product order (FPO) has increased from 3925 in 1991 to 5293 in 2000 with a composite growth rate of 34.85 per cent.

Reimposition of union excise duty on fruit products in the union budget of 1997-98 had hindered the growth of this industry resulting in a meagre expansion of the installed capacity but also in increasing the idle capacity, thereby increasing the cost of products for the consumer[4]. It resulted in a decline in production from 9.6 lakh tonnes in the year 1997 to 9.1 lakh tonnes in 1998. The adverse effect was visible in the year 1999, 2000 and 2001.

Marketing of Fruit Products

It has been reported that the domestic consumption of value added fruit products is low compared to the primary processed food in general and fresh fruits in particular, which is attributed to higher incidence of tax and duties including that on packaging material, lower capacity utilisation, non-adoption of cost effective technology, high cost of finance, infrastructural constraints, inadequate farmer processors linkage leading to dependence upon intermediaries.

The present marketing practice involves a number of intermediaries such as commission agents, whole sellers, sub-whole sellers, retailers etc. resulting in the grower realizing hand by 30-40 per cent of the price paid by the consumer. The balance is shared by the intermediaries at different stages including a certain percentage of system wastage. Thus the consumer is paying two and a half to three times the price which the grower is actually realizing. There is, therefore, a distinct need for increasing the growers' share and evolving a suitable marketing system where both growers and consumers are benefited. One of the solutions could be organization of growers cooperatives societies for marketing, including retailing of the produce to avoid intermediaries.

Role of Cooperatives in Marketing of Fruits

There are about 12 state/central level societies and 275 primary marketing societies directly engaged in marketing of fruits in India[5]. However, there is no exclusive federation of fruit cooperatives societies at the national level, for marketing of horticultural produce. This work is being undertaken by the National Agricultural Cooperative Marketing Federation of India (NAFED) in collaboration with affiliated State Federations and Primary Marketing Societies. NAFED handles a number of commodities like apple, grapes, mango, orange etc. The cooperatives in marketing of horticultural produce mainly fruit items during the last three years, is given in Table 3.4.

Table 3.4
Fruits marketed by cooperatives

Year	*Value (Rs. in crore)*
1992 - 93	276.63
1993 - 94	362.10
1994 - 95	437.90

Source : Productivity Vol. 36, No. 4, January - March, 1996, p. 552

Export of Fruits and Processed Fruits

The main fruits that enter the export market are mangoes, grapes, apples, citrus, bananas, sapota, litchis etc. The total exports, according to the MFPI, Annual Report 2001-02, are about 50 thousand tonnes valued at Rs. 600 million. The main destinations being Middle East, U.K., Europe and to some extent Singapore, Malaysia etc. By the end of the century it is targeted to increase the export to 355 thousand tonnes valued at Rs. 6,940 million.

India's major exports include fruits pulps and juices, pickles, chutneys, concentrated pulps and juices, and frozen fruits. The export values of processed fruits has increased from Rs. 194 crore in 1991-92 to Rs. 1345 crore in 2000-01 as indicated in Table-3.5. With the ongoing modernization, technology upgradation as well as fresh investment in this sector accompanied with tie ups with the world leaders, it is expected that the export of Indian processed foods will record a major growth in the coming years.

Table 3.5

Exports of Processed Fruits

Year	*Value (Rs. in crore)*	*% Growth*
1991 - 92	194.00	—
1992 - 93	263.00	35.6
1993 - 94	318.00	20.9
1994 - 95	409.00	28.6
1995 - 96	479.00	17.1
1996 - 97	473.77	(1.1)
1997 - 98	761.50	60.7
1998 - 99	705.60	(7.3)
1999 - 00	993.60	40.8
2000 - 01	1345.50	35.4
2001 - 02 (target)	1265.00	—

Source : Annual Report 2001-02, Ministry of Food Processing Industry, Govt. of India.

Figure in bracket indicates negative percentage growth.

The percentage growth of export of processed fruits has revealed a positive trend except for the year 1996-97 and 1998-99. Negative growth rate is highest with minus 7.3 per cent during 1998-99. The reason could be traced back to the decline of production of processed fruits in 1998 due to reimposition of union excise duty on this items[6]. It is observed that export trend of processed item will continue to grow. The composite growth during 1991-92 to 2000-2001 was calculated to be 593.5 per cent.

PART - II

Production and Marketing of Fruits and Processed Fruits in North Eastern Region

The entire North Eastern Region is endowed with fertile soil, high rainfall and a favourable climate for growing different kinds of fruits besides varieties of other crops. Horticulture could be one of the strongest base of the economy of the North Eastern Region of India, if properly controlled the post harvest losses.

Owing to physical, economic and natural factors the region could not make much headway in the sphere of growth performance and development compared to other states of the country. It is important to accept the fact that the data base on horticulture in most of the North Eastern states is rather poor. Unfortunately, dependable and precise statistics of area, production and productivity of many fruits are not readily available and that is why the horticultural development do not get correctly as projected, resulting in handicaps in perspective planning and development.

On a rough estimate, the area under fruit crops in this region is about 209.37 thousand hectares with an estimated production of 2,127.13 thousand tonnes of fruits. Among the different varieties of fruits, production of banana rank first with 751.31 thousand tonnes followed by pineapple with 480.40 thousand tonnes, jack fruits with 378.83 thousand tonnes and oranges with 149.79 thousand tonnes which has been mentioned in Table 2.1 of the previous chapter.

Processing of fruits has a special significance in the North Eastern Region because of the high rate of spoilage on account of lack of storage facilities and inaccessibility to far flung market centres. The major processed items in the region include juice and pulp, jam and jelly, pickles, passion fruits, ready to serve beverages, synthetic syrups, squashes and the like. The total installed capacity of fruits processing industry in the region comes to about 502 metric tonnes annually [7].

The North Eastern Council (N.E.C.) and the Food Processing Ministry at the centre have formulated a number of measures to improve processing and marketing of the products. Out of the current production of fruits in the region about 30-40 per cent of fruits are being spoiled annually[8]. Therefore, there is a vast scope for processing of these products to value added products to earn higher revenue. In fact, a bulk of these products can also be exported to earn valuable foreign exchange. The N.E.C. also has formulated a regional project on marketing of agro-horticultural products for the region. Various aspects of the project include setting up a multi-commodity cooperatives, creation of processing facilities and sales promotion support to marketable surplus and processed products. In the early part of the year 2000, there were 61 fruit processing units functioning in the region. The centre has provided financial assistance under the plan schemes to processing units of this region. Some more processing units are also coming up in next few years which can be seen from Appendix-III.

Marketing performance of the processing units of this region, is yet to be developed fully. The unit gives less importance to market survey and research as indicated in the questionnaire that almost all the units of this region has responded of not conducting market survey before and after the commencement of their business. This indicates that, they produce what they could produce and whatever they like and not according to the demand in the market. That is why, their product could be sold in the local market alone and only few products are tested outside the state and in abroad. But it is learnt that sometimes the product of the region sold out particularly in foreign countries were returned, the reason may be due to non complience of quality, and other sub standard detected by them. The detail about this reason could not be specified exactly.

Some Basic Issues Pertaining to Marketing in North Eastern Region

There are certain inherent problems in developing marketing in North Eastern Region which need immediate attention.

Transport Network

The entire North Eastern Region is geographically isolated and linked with the rest of the country by an narrow corridor of about 56 km [9]. By and large the entire region, except the Brahmaputra valley, is a hilly and mountainous terrain and economically underdeveloped compared to the rest of the country. 'The region has a total road length of 1.33 lakh km[10]. This includes about five thousand km. length of National Highway connecting all state headquarters and other urban centres in North Eastern Region'. As of now, only half of the total villages in the region are connected by all weather roads, while the remaining villages are without proper roads. The total road length per 100 sq. km. of Tripura is 140.4 per cent, of Assam is 86.8 per cent and that of Nagaland is 82.8 per cent against 73 per cent of the country as a whole, but the percentage of the whole North Eastern Region is just 52.08 which is less than the all India average as the percentage of the other state in the region are comparatively low, as it is shown in the Table-2.3 of the previous chapter.

Market Organisation and Structure

Markets are poorly developed in the entire North Eastern Region. At present there does not seem to be any uniform pattern of organisation and management set up of primary and secondary markets. It was observed that it differs from state to state, district to district and market to market. But also, there are different levels of marketing organisations units exists in this region. They are, private level, state level, regional level, national level and other marketing organisations as shown in the Table-3.6 below.

Table - 3.6

Marketing Organisation in the North Eastern Region

State.	*National level*	*Regional level*	*State level*	*Private level*	*Others*	*Total marketing organisation*
Arunachal Pradesh	3	1	8	30	NIL	42
Assam	17	6	14	31	38	106
Manipur	3	1	6	14	3	27
Meghalaya	3	4	11	29	9	56
Mizoram	2	1	7	20	2	32
Nagaland	2	1	8	15	3	29
Tripura	3	1	10	18	5	37
Total	33	15	64	157	60	329

Source : North East India Data Bank, NIC Assam State Unit Guwahati.

Private Level Marketing Organisation

The private level marketing organisation renders services in awareness studies , marketing strategy formulation, feasibility studies, market survey for agro-products, distribution of consumer products, exporter and importer, and advertising and sales promotion etc.

State Level Marketing Organisation

The state level marketing organisation helps in marketing of local products, policy implementation on Government marketing schemes, exploring export market for local products, organising exhibitions and trade fairs and in organising training programmes on marketing etc. It also gives aids, advice and assistance on marketing of local products.

Regional Level Marketing Organisation

The regional level marketing organisation promotes

services in projects consultancy providing raw materials, marketing of fruits and in various training programmes on processing units for procuring critical items and sponsors research studies on marketing etc.

National Level Marketing Organisation

The national level marketing organisation helps in counselling on marketing, providing sub-contract exchange facilities and gives assistance to integrated units involving production and marketing of horticultural products. It renders services in issuing of export and import licence, marketing and inspection of food products (AGMARK) and assessing in getting and execution of orders under central government stores purchase programme.

Other Marketing Organisation

Apart from the above mentioned organisations there are 60 other marketing organisations which comprises marketing services, export marketing services, commission agents, advertising and publicity and associations etc. This marketing organisation renders services in counselling on export marketing and in marketing of local products.

The region has a total of 329 marketing organisation as indicated in the Table - 3.6, of which 33 are national level, 15 are regional level, 64 are state level, 157 private level and the rest 60 are categorised as others.

In the national level marketing organisation there are total 33 organisations of which 17 are in Assam, 3 each in Arunachal Pradesh, Manipur, Meghalaya and Tripura and 2 each in Mizoram and Nagaland respectively.

In the regional level marketing organisations there are total 15 organisations. Of which 6 are in Assam, 4 in Meghalaya and 1 each in Arunachal Pradesh, Manipur, Mizoram, Nagaland and Tripura respectively.

Out of the 64 state level marketing organisation 14 are in Assam, 11 in Meghalaya, 10 in Tripura, 8 each in Arunachal Pradesh and Nagaland, 7 in Mizoram and 6 in Manipur.

In the private level marketing organisation there are total of 157 organisations. Of which 31 are in Assam, 30 in Arunachal Pradesh, 29 in Meghalaya, 20 in Mizoram, 18 in Tripura, 15 in Nagaland and 14 in Manipur respectively.

In others group, out of the total 60 marketing organisations 38 are in Assam, 9 in Meghalaya, 5 in Tripura, 3 each in Manipur and Nagaland and 2 in Mizoram.

In most of the markets it is the local autonomous administrative bodies and private persons who are the controlling authorities. The local bodies who control the rural markets consider the markets as an earning place rather than a service institution. In some areas, "Bazar Samitis" or Market Committees have been consisting of the local elders or "Gaonburas", who look into the aspects of allotment of shops, fixing rents, settlement of disputes etc. The function such as regulation or trade practices, development of markets by providing basic infrastructure and services etc. are not given any importance. The markets in the remote hilly areas are still in a very rudimentary shape.

It is also worth mentioning that central assistance is available for development of the markets to the extend of fifty per cent of the project cost limited to Rs. 15 lakh in case of terminal markets for fruits, Rs. 5 lakh in case of whole sale assembling markets and Rs. 1.5 lakhs in case of rural markets, provided the markets are regulated or institutionally linked with regulated market. This has been availed by the state of Assam and Tripura[11]. Moreover, these local bodies have control over the marketing process except collecting levies and tolls.

Enactment of Market Legislation

In India, agricultural produce market act are enacted in different states basically to safeguards the farmers in getting a fair deal. However, the process of enactment and enforcement of these Acts is not uniform all over the region. In Assam, the Assam State Agricultural Produce Market Act, 1972 was implemented in 1976. But enforcement of the Act in the state is very poor. No transactions is practically done in any yard. The market committees are collecting market cess and the rural markets are still in the hands of the local bodies.

Tripura is the first state to extend the Bombay Agricultural Produce Market Act, 1939 in 1956. Later on, Tripura Agricultural Produce Market Act was passed in 1980. The Meghalaya Agricultural Produce Market Act, 1980 has been enacted in Meghalaya. However, not much progress has been made in enforcement and notification of markets. Nawlong near Shillong is the first market to be regulated in Meghalaya.

In Manipur, the Bihar Agricultural Produce Market Act, 1960 was extended in the state in 1963. The state has not made any progress in enforcing the Act. In the state of Nagaland the Bill has been enacted and for the state of Mizoram and Arunachal Pradesh the draft bill is under process. No state of the North Eastern Region has separate Directorate of Agriculture Marketing for market administration. The state Agricultural Marketing Board usually control the function of market committees.

Storage Processing and Transport Communication

In the entire region, storage facilities in rural markets and primary whole sale assembling markets practically do not exist. The F.C.I. (Food Corporation of India) and C.W.C. (Central Warehousing Corporation) have built up a storage capacity to the extend of 73.59 lakh in the public sector. The region has only 10 cold storage against the all India total of 452 warehouses with an installed capacity of 86,000 MT which are under utilised[12]. There are no community storage in the producing areas. Recently, Government of Assam have started a project to establish cold storage in the regulated market yards for fruits and vegetables[13]. Moreover, transport bottlenecks are disrupting supplies in the hilly region. As compared to 73 km. per 100 sq. km all India average, this region has only 52.08 km of roads per 100 sq. km[14]. The 'farmers have to walk more than 20 km. to reach a market place as against 5 km in Punjab. It is reported that the quantum of produce on head load to the primary market is 88, 70, 56 and 48 per cent respectively in Manipur, Meghalaya, Assam and Mizoram'[15]. The common means of transport are shoulder load, head load, back load, push and pull cart, bullock cart, bicycles, public and private bus and short distance trains.

Market information is also inadequate in the region. Village traders, commission men visiting the rural areas are the only source of market information. Regular broadcasting of market intelligence service through print and electronic media is not so effective in the region.

Post Harvest Cares

The post harvest care like sorting, grading and standardization, packaging etc. are not common in this region. Factor like degree of maturity, damaged ones mixed with better ones, undersized, ripen and unripened through faulty measures are common in this region. So also matured and immature fruits are harvested together including damaged and diseased marketed which will have poor shelf life in storage and long distance transport. The losses in transit is estimated at 35 to 51 per cent in case of some fruits[16].

Domestic Market

Absence of domestic market for fresh and processed fruits and vegetables is an important consideration. Fruits like banana, pineapple etc. are supplied and find their way to adjoining states. Mizoram finds it hard to absorb all the ginger produced and as a matter of fact losses are being incurred which are being subsidised by the Government.

Horticultural Activities in Whole Farm Business

Horticultural crops constitute the major part of homestead

farming. In general every rural household is having some fruit trees, plantation crops and a small kitchen garden for growing seasonal vegetables. Of course, some selective households may have well laid out orchard. Homestead farming is an important part of the whole farm business of a farm. It plays important role in supplementing the farm business income beside providing employment to human labour and utilisation of other stock and flow resources. Field study conducted by the researcher of the Department of Agricultural Economics, Assam Agricultural University, Jorhat worked out that of the total area, employment, working capital utilisation and realisation of net project in the whole farm business, the relative share of homestead farming were 6.81, 6.75, 28.69 and 35 per cent respectively in the tribal farm and 6.39, 6.46, 33.04 and 34.96 per cent in the non-tribal farms[17]. Similar results have also been reported in many other studies carried out in different districts of the states.

Present Status of Marketing of Horticultural Produce

Horticultural produce which comprises mainly fruits and vegetables needs a sound marketing system for its development. Primarily being perishable in nature, a poor marketing system may adversely affect the farm economy. Low producer price vis-a-vis high consumer's price, heavy wastage in transit, deterioration in quality, spatial and temporal disequilibrium affecting market demand and supply over time and space make the market inequitable.

In rural areas, usually farmers assemble the marketable surplus in the road side or in any open space where the traders from town come and procure the produce. During seasons, such 'hat' are held daily. However such assembly centres are also inaccessible in the rural areas by all categories of farmers. Some farmers try to arrange the primary sales of their produce directly at town. Group of farmers also bring their produce to the town by renting public carriers. The village traders also collect the produce in the village 'hat' carries it to the bus point, retails in the town and returns. As the farmers are unorganised, unskilled and untrained compared to organised and skilled traders their competition becomes imperfect, affecting farmer's price. Inter state and inter district movement of fruits and vegetables are done by the district wholesalers or by their commission men.

In an subsistence agriculture like ours, poor economic status of the farmers, lot size of their produce, poor transport and communication, seasonality, perishability etc. compels them to sell to the market middlemen who are playing havoc in the market, create artificial market environment with an adverse impact on rural economy. In fruits and vegetables marketing, such institutions are multiplied to itinerant village traders, whole salers both local and distant, commission men, pre harvest contractors, processors, packers, collers, truckers etc. However, the pre harvest contractors are playing an important role in fruits marketing particularly for citrus marketing in Assam. These contractors also act as money lenders cum traders. They render all kinds of financial help during lean period and purchase the garden in advance. In such contracted sales, the gardens are not only deteriorated but it also affects the producer's price drastically. In some cases, about 91 per cent of total sales at farm level are done through this contracted sale earning, about 41 per cent of total margins in the channel (Gogoi, 1978). The producers could get only 17.85 per cent of consumers' rupee[18].

However, no government agencies are taking the responsibility of concentration, transformation and equalisation for fruits and vegetables for substituting contractors rendering all possible financial help. The role of NERAMAC (North Eastern Regional Agricultural Marketing Cooperation Limited) is insignificant in the state. The state does not have fully developed producer's marketing co-operatives societies for fruits and vegetables.

The market education and research in the country have been developed in the management institutes and agricultural universities of the country. Assam Agricultural University is the only institute in the entire North-East India in this respect.

Other research organisations in North East India like ICAR (Indian Council of Agricultural Research) Research complex, National Productively Council and Agro Economic Research Centre for North East India also conduct some cross sectional studies in the region. Assam Agricultural University has conducted several studies on Agriculture marketing through students' research. The research conducted in this university are mainly related to marketed and marketable surplus, channels of distribution, market middlemen disposal, pricing efficiency, storage and price policy for various commodities. Besides these, there are some ad-hoc schemes and non-project faculty research and agricultural marketing.

PART III

Viability and Suitability of Horticulture Business in Manipur

The researcher belong to the state of Manipur and since one of the objective of the present study is to identify the viability and suitability of fruit processing industries in the hill districts of Manipur, the attempt is made in this section.

The varied agro-climatic condition and rich agricultural resources viz. rain, water, vegetation etc. of Manipur gives a great potential for the development of horticultural crops. Tropical fruits like pineapple, orange and other citrus and temperate fruits like pear, peach etc. are suitably grown and have great potentiality. This is mainly in the hills which constitute 9/10th of the states total geographical area. Off season vegetables like green pea, cabbage and other leafy vegetables are also grown in the hills. Floriculture is another aspect which has tremendous export potential but not exploited for economic development of the state. Manipur is very fast emerging as a force in plantation crops mainly fruits and vegetables, tea, coffee and rubber. Spices also have a very good potential. Yet another presently unexploited but with great potentiality is the aromatic and medicinal plants in the state. All these horticultural crops put together can transform the economy of Manipur from the present low level to a very high level by exploiting both in the domestic as well as the international market. Horticulture in Manipur can rightly be called "The future money earner of the state". Since the scope of plantation of fruit crops in the plain area is rather limited, the major thrust will have to be on vegetables and spices.

The systematic expansion of horticultural crops in the hilly region will reduce the pressure on jhuming or shifting cultivation which is at present a very popular form of cultivation in the hilly region.

Fruits, vegetables and spices are the important crops which contributes considerable shares in the economy of the state. Fruits are grown mostly on the hill slopes in tribal areas and the contribution to the tribal economy of the state is quite conspicuous. As mentioned earlier, fruit growing in the hills will help in controlling the jhum cultivation. Of the vegetables, cabbage, cauliflower, tomato, beans, cucumbers, brinjal, bhindi, squash, potato etc. are the crops that help the farmers economically and more sound for those who are inhabiting along the river valleys. Considerable quantities of various types of vegetables, particularly rabi vegetables are marketed outside the state and other parts of the country, thereby improving the economy of these farmers. Chilly and ginger are important spice crops that add to the economy of the growers.

The gross value of fruits and vegetables produce in Manipur for the year 1993 is worked out to be Rs. 49.90 crore, of which fruits alone accounted for Rs. 33.15 crore and vegetables accounted for Rs. 16.75 crore[19].

In terms of contribution of different fruits and vegetables to the total gross value, it is observed that pineapple accounts for 36.79 per cent and banana for 35.38 per cent ranking first and second respectively as can be seen from the below Table - 3.7 and Table - 3.8. Among the vegetables, tomato accounts for 36.63 per cent, cabbage for 23.91 per cent and cauliflower for 22.75 per cent ranking first, second and third respectiely as it is shown in the below Table-3.8

Table - 3.7

Fruits' Contribution to Total Gross Value

Fruit items	*Contributon to total gross value*	
	Value (in crore)	*Percentage*
Pineapple	12.20	36.79
Banana	11.73	35.38
Lime/Lemon	1.54	4.65
Orange	1.48	4.48
Others	6.20	18.70
Total	**33.15**	**100.00**

Source : *Techno-Economic Feasibility Study for Development of Horticulture in Manipur (Survey Report of National Horticulture Board, 1994) p. 37*

* Other fruits includes papaya, pear, peach, plum, jackfruits, guava, mango, etc.

Table - 3.8

Vegetables' Contribution to Total Gross Value

Vegetable items	*Contributon to total gross value*	
	Value (in crore)	*Percentage*
Tomato	5.97	35.63
Cabbage	4.01	23.91
Cauliflower	3.81	22.75
Pumkin	1.34	8.02
Others	1.62	9.60
Total	**16.75**	**100.00**

Source: *Techno-Economic Feasibility Study for Development of Horticulture in Manipur (Survey Report of National Horticulture Board, 1994) p.37.*

* Other vegetables includes brinjal, bhindi, squash, arum etc.

Looking into the pineapple map of Manipur (Map No. 1), Bishnupur district and Thoubal district has the highest pineapple production in the valley. In the hills, Churachandpur district and Senapati district has produced a large quantity of pineapple which can be commercially exploited. Mention may be made that till now officially there are only 13 fruits processing units in Manipur, of which 12 units are within Imphal itself and the remaining one unit is at Thoubal (see Appendix - II). There is a new project proposal named 'Khongjom pineapple processing industry, Khongjom, Tekcham' in Thoubal district. This was already forwarded to the Ministry of Food Processing Industries, Government of India and is now under consideration (see Appendix - III). In Bishnupur district, a new project called 'Bishnu fruit and vegetable plant' in the heart of Bishnupur bazar has been approved and sanctioned the sum of Rs. 11,20,000/- by the Ministry of Food Processing Industries, Government of India (see Appendix - III). Also it has been reported that building construction of the said project has been completed but the machinery are yet to be installed. But the similar project are yet to be take up in Churachandpur and Senapati, the hill districts of Manipur.

The three hill districts, namely, Tamenglong, Churachandpur and Senapati has the great potentiality for producing banana in Manipur as can be seen from the Banana Map of Manipur (Map No. 2). But none of the projects for processing banana are yet to be set up in the state particularly in the hill district of Manipur. Other than generating income and employment generation, setting up of fruit processing unit in the hill district of Manipur will definitely help the farmers who are not only selling their product at the throw away price but very often could not find their surplus product which is allowed to be rotten in the field[20]. This remark is justified for all types of fruits.

Lime and Lemon Map of Manipur (Map No. 4) reveals that Churachandpur, Tamenglong and Ukhrul, all the three hill districts has a great potential for producing lime and lemon in Manipur. Tamei Area Fruit and Vegetable Growers cum

Map No. 1

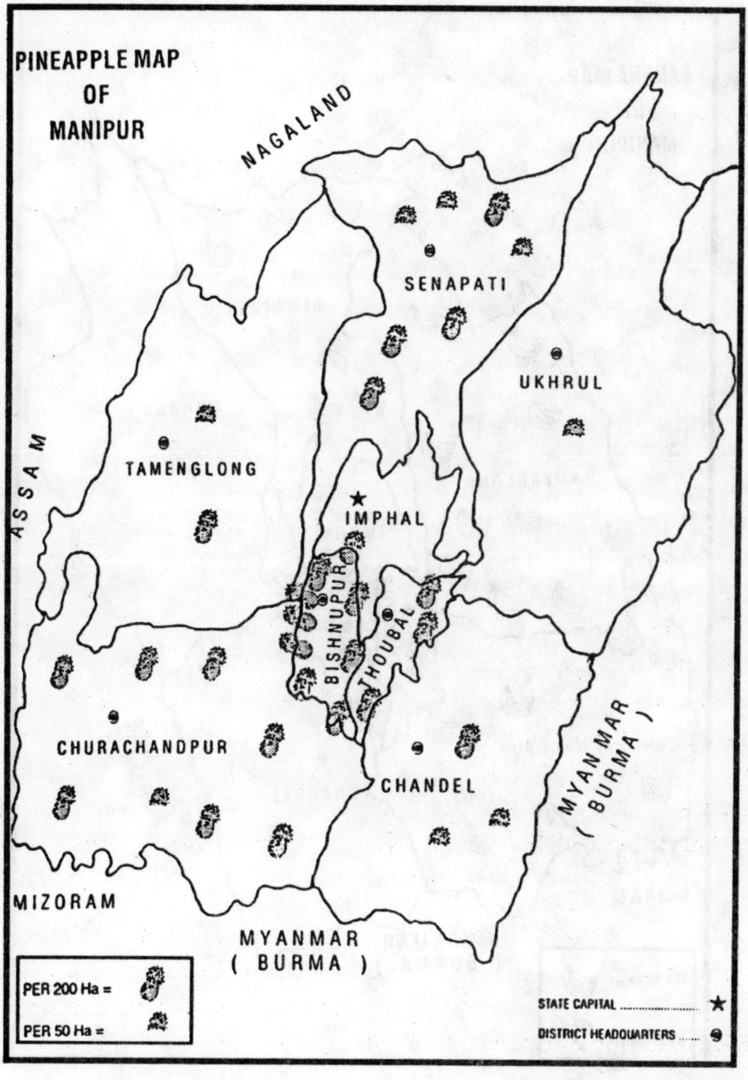

Source : Techno-Economic Feasibility Study for Development of Horticulture in Manipur, National Horticulture Board, Survey Report, 1994.

Map No. 2

Source : Techno-Economic Feasibility Study for Development of Horticulture in Manipur, National Horticulture Board, Survey Report, 1994.

Marketing Cooperative Society Ltd., a new project in Tamenglong district has already been approved and got a sanction of Rs. 6 lakh from the Ministry of Food Processing Industries, Government of India. In Ukhrul district, a new project called 'PARDA, Lemon Processing Plant, Kachai, Ukhrul' has been approved and has already got a sanction of Rs. 8.51 lakh from the Ministry of Food Processing Industries, Government of India (See Appendix - III). Mention may be made that a particularly variety of lime called Kachai lime or K-lime[21] in short which is of superior quality is found at Kachai village in Ukhrul district and its adjoining villages, where this lemon processing plant is installed.

Looking in the orange Map of Manipur (Map No. 3), it is found that Tamenglong district is the most potential district in Manipur for growing orange. But the oranges are available in plenty during the season from different corners of Manipur. It is learnt from the FPI section of the Directorate of Commerce, and Industries, Government of Manipur that Grihang village which is near the Myanmar border in Kamjong sub-division of Ukhrul district and its adjoining villages has a great potential for growing superior quality of oranges. The growers even brought oranges from this area and marketed them at the district head quarter Ukhrul and in Imphal market. The yearly production is roughly estimated to be around 13,000 tonnes[22].

Passion fruit is another commercial crop which can be cultivated in Manipur. It has a longer shelf life comparing with other fruits and has export market potential. A Mao based NGO called Good Samaritan Social Service Association has taken up a giant project on passion fruit cultivation in Senapati district, which is the first of its kind in the state. The association has been promoting a large number of local farmers by providing them technical know-how on organic passion fruit farming and is looking for export of fresh passion fruits in the Western Countries particularly USA in the coming year. The association is asking for financial assistance from the Ministry of Food Processing Industries, Government of India (see Appendix - III).

Map No. 3

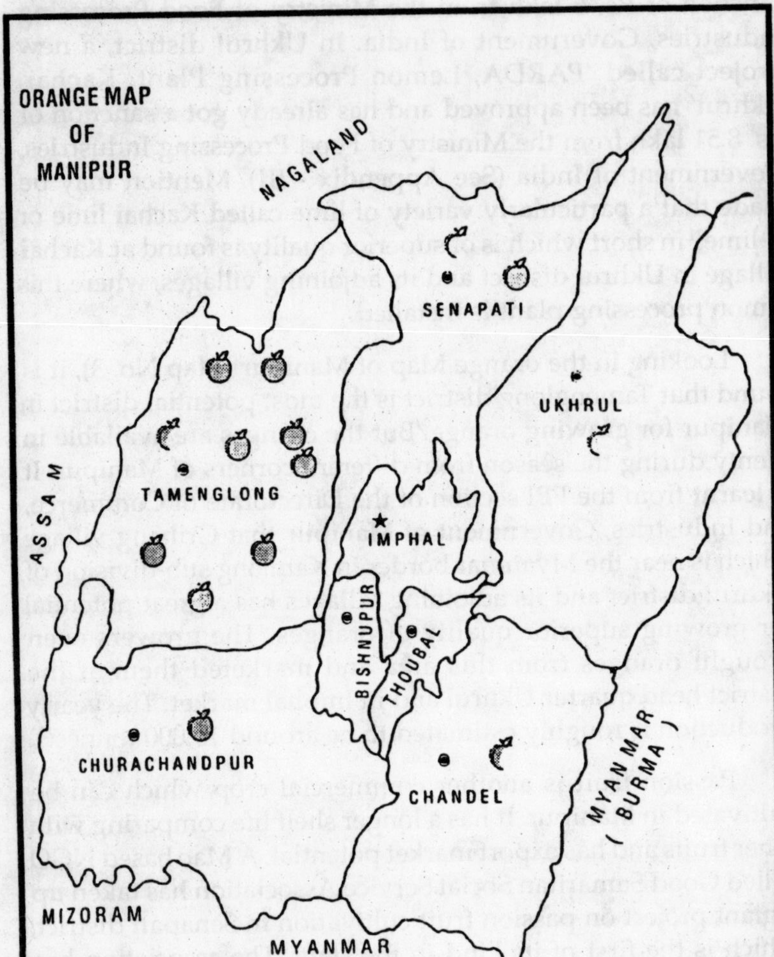

Source : Techno-Economic Feasibility Study for Development of Horticulture in Manipur, National Horticulture Board, Survey Report, 1994.

Map No. 4

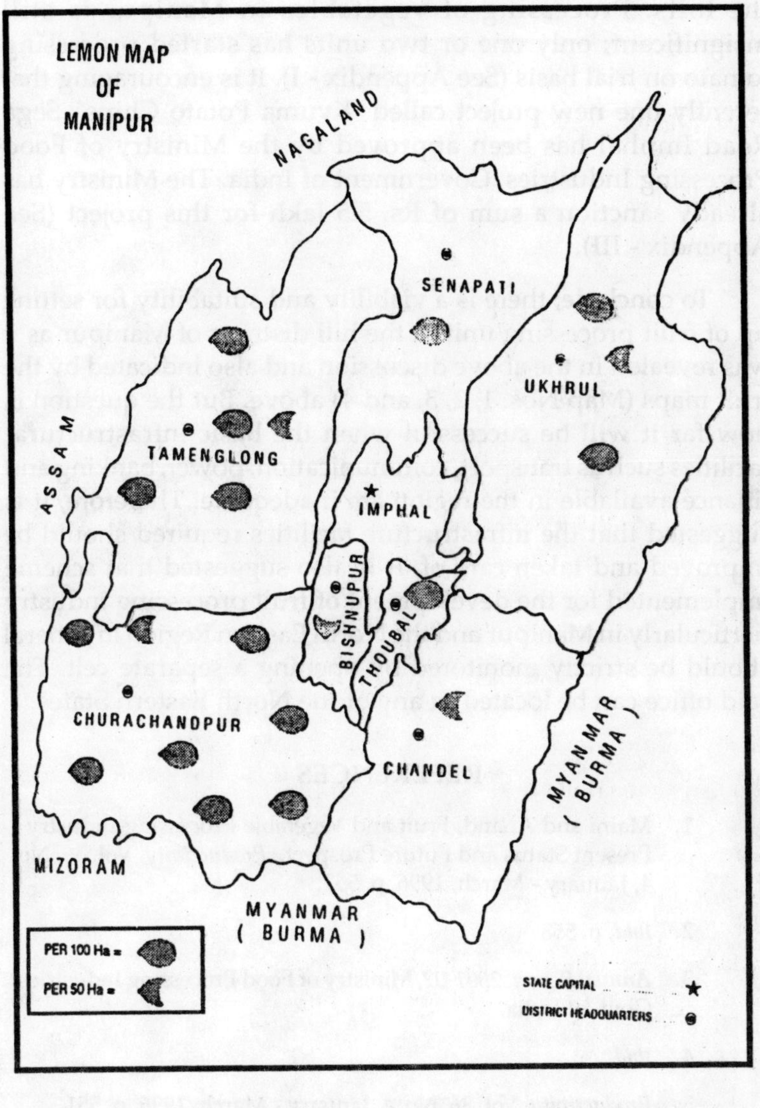

Source : Techno-Economic Feasibility Study for Development of Horticulture in Manipur, National Horticulture Board, Survey Report, 1994.

Vegetable crops specially tomato, cauliflower can be extensively grown in the valley. Cabbage can also be grown in the hills. Processing of vegetables in Manipur is still insignificant; only one or two units has started processing tomato on trial basis (See Appendix - I). It is encouraging that recently one new project called 'Kyuma Potato Chips', Sega Road Imphal has been approved by the Ministry of Food Processing Industries, Government of India. The Ministry has already sanction a sum of Rs. 5.5 lakh for this project (See Appendix - III).

To conclude, there is a viability and suitability for setting up of fruit processing unit in the hill districts of Manipur as it was revealed in the above discussion and also indicated by the fruit maps (Map Nos. 1, 2, 3, and 4) above. But the question is how far it will be successful when the basic infrastructural facilities such as transport, communication, power, banking and finance available in the region are inadequate. Therefore, it is suggested that the infrastructure facilities required should be improved and taken care of. It is also suggested that scheme implemented for the development of fruit processing industry particularly in Manipur and the North Eastern Region in general should be strictly monitored by opening a separate cell. The said office can be located in any of the North Eastern States.

REFERENCES

1. Maini and Anand, Fruit and Vegetable Processing Industry - Present Status and Future Prospect. *Productivity*, Vol. 36, No. 4, January - March, 1996. p. 557.
2. *Ibid*, p. 558
3. *Annual Report, 2001-02*, Ministry of Food Processing Industries, Govt. of India.
4. *Ibid*.
5. *Productivity*, Vol. 36, No. 4, January - March, 1996, p. 551.
6. *Annual Report, 2001-02*, Ministry of Food Processing Industries, Government of India.

7. Press Information Bureau (PIB) : Food Processing in North East, the *Assam Tribune*, March 14, 1994.

8. Singh, R.K. P., et al, Post Harvest Management of Fruits and Vegetables in North Eastern Region of India. Paper presented during the National Seminar on Management Stratigies for North Eastern Hilly Ecosystem, Shillong : August 8-19, 1998.

9. Borthakur, D.N. : Agriculture of the North Eastern Region with Special Reference to Hill Agriculture. p. 1.

10. Basic Statistics of North Eastern Region 2000, NEC Publication, Shillong. p. 110.

11. K.C. Talukdar and B.C. Bhowmick, *Marketing of Perishable Products*, BR Publishing Corporation New Delhi, 1993.

12. Basic Statitics, p. 80.

13. Bhowmick, B.C. : Marketing of Fruits and Vegetables in North Eastern Region Paper Presented in the Training Organised by NIAM, Jaipur on Agricultural Marketing Practices in North Eastern Region. Dec. 6-9, 1994. p. 13.

14. Basic Statistics, *op. cit.,* p. 110

15. Bhowmick, *op. cit.,* p. 14.

16. *Ibid.*

17. *Ibid*, p. 17.

18. *Ibid*, p. 17.

19. Techno-Economic Feasiblilty Study for Development of Horticulture in Manipur, a Survey Report of National Horticulture Board, 1994. p. 36.

20. Singha, Rajmani and Yarso, A.S. : Agro-based Industries for Food Security in Manipur. Paper Presented on National Seminar on Food, Nutrition, Food Security, Public Distribution System in North - East India at Shillong (27 - 28 November, 2000).

21. Recommended Varieties of Fruit Crops : Technological Options for North Eastern Hill Region ICAR Research Complex for NEH Region, Barapani, Meghalaya, 1999.

22. Avai Horam : A Brief Note on an Industrial View of Ukhrul District Specially in Food Processing Sector. Seminar Paper Presented on Development of Food Processing Industries in Tribal Areas of Manipur. Organised by Directorate of Commerce & Industries which was Sponsored by MFPI, Government of India. Held on 12th March, 1997 at Ukhrul.

4

Performance Evaluation of Fruit Processing Units

Introduction

In the foregoing chapters we have already discussed the present status, production and marketing trend of fruit processing units in the North Eastern Region and in India in general. The present chapter deals with the performance evaluation of fruit processing units of the region. Before evaluating the performance of the units, let us know the brief background of the fruit processing units selected for the study.

PART - I

Fruit Processing Unit Profile

Public Sector : Under the public sector, three fruit processing units have been selected for study. They are :-

1. Govt. Fruit Preservation Factory (MAGFRUIT), Manipur

The Govt. Fruit Preservation Factory is located in the campus of Agriculture and Horticulture complex at Mantripukhri, Imphal on the northern side of the National Highway 39. The factory was started in the year 1958 for the purpose of demonstration and to impart training to the public for better utilisation of the Horticultural crops. It is being run departmentally under the Directorate of Horticulture and Soil Conservation, Government of Manipur. The unit is marketing their product under the brand name "MAGFRUIT" with the FPO Licence No. 1443. At present, the factory employed altogether 36 employees, 23 of them are full time employees and 13 of them are part time employees. The main items of products produced are squashes of pineapple, orange and lemon.

2. Fruit Juice Concentration Plant, Mizoram

The fruit juice concentration plant, is located at Chhingchhip, MIFCO Ltd. complex, Aizawl, Mizoram, This unit is manufacturing and marketing their products under the brand name 'MIFCO' with the FPO Licence No. 11369. It has started processing of fruits since 1998 with the main objective of promoting food processing in Mizoram. The unit has 25 full time employees and part time workers are employed depending on the work load. Juices of passion fruits, pineapple and orange are the main products of this unit.

3. Govt. Fruit Preservation Factory, Meghalaya

In Meghalaya, the most popular brand name 'MEG' with FPO Licence No. 653 is manufactured and marketed by the govt. Fruit Preservation Factory, undertaken by the state government. The unit is located at Fruit Garden near the Directorate of Horticulture, Government of Meghalaya. This unit has started processing of fruits since 1955. The main objectives of this unit are to preserve and prevent spoilage of perishable surplus of fruits grown abundantly in the state and to create a ready made market for farmers and fruit growers. The unit has 15 persons as full time workers. The main items produced by this unit are the juices of orange, sohiong (local name) and pineapple and the squashes of sohiong and orange.

Co-operative Sector : Under the cooperative sector another 3 fruit processing units were taken for study. They are :

1. Manipur Fruit Processing and Cold Storage Co-operative Society Ltd., Manipur

The Manipur Fruit Processing and Cold Storage Co-operative Society Ltd. is located at Kombirei Road, Lamphelpat near the office of the Deputy Commissioner, Imphal West District. This unit is runned co-operatively and traded under the brand name 'Heirang' with the FPO Licence No. 4011. The factory was established on 24th February 1975 with the objective of helping the growers with better prices by eliminating the role of middle men and also to enable people to get fresh and processed fruits at favourable prices through out the year. The unit has 12 full time employees at present. But part time employee (s) are engaged from time to time with the number fluctuating between 5 to 12. The main products produced by this unit are the squashes of orange, pineapple and lemon.

2. WAIFRUIT (Agrotech and Aqua products)

The WAIFRUIT (Agrotech and Aqua-products) is located at YVU Bhawan, Indo-Burma Road, Thoubal Wangmataba, Thoubal. The unit is under co-operative sector and has started processing of fruits from 1st August, 1995. The unit is manufacturing and marketing their product under the brand name 'WAIFRUIT' with the FPO Licence No. 9724. The main objectives of this processing unit are to generate employment opportunities and to generate income mainly from outside the state by selling their product and also to produce good quality diet. The unit employed altogether 53 persons, of which 11 of them are full time workers and 42 workers are part timers. Squashes of orange, pineapple and lemon are the main products of this unit.

3. Modern Food Processing Co-operative Society Ltd., Assam

The brand name "NIRU'S" with the FPO Licence No. 7861 is manufactured and marketed by the M/S Modern Food

Processing Co-operative Society Ltd. The processing unit is located at 'NIRUPAMA' Chenikuthi Hill side, Guwahati, Assam. It had started processing of fruits under co-operative sector from June 1990. The unit has 14 employed persons and all are full time workers. The main objectives of this processing unit are to process indigenous fruit and to create awareness of food processing to the public and to generate employment opportunities to the people of the state. The unit produce mainly the squashes of carombola and olive pickle.

Private Sector : Eight fruit processing units were selected for study under private sector. They are :

1. Ratna Fruit India, Manipur

Ratna Fruit India is located at Uripok Polem Leikai, Imphal (west district). It is under private sector, set up with the main objectives of self-employment and socio-economic upliftment of the state. The unit was commissioned for processing in the year, March 1992 under the brand name 'RATNA' with the FPO Licence No. 7779. There are 6 full time employees employed by this unit. The chief items produced by this unit are the juices of orange, pineapple and lemon and the squashes of orange and pineapple.

2. Manipur Hill Fruits Processor, Manipur

The brand name 'CHANDNI' with the FPO Licence No. 8320 is manufactured and marketed by the M/S Manipur Hill Fruits Processor which is located at Pungdongbam, B.P.O. Yourbung, P.O. Lamlong, Imphal east district. The unit started processing fruits from 20th February 1991. This unit is under private sector giving employment opportunity to 7 full time employees and 6 part time employees, together 13 persons are employed. Employment and income generation are the main objectives of this unit. The main items of products produced by this units are the squashes of pineapple, orange and lemon and the lemon pickles and soft drinks made out of these fruits.

3. Thangjam Agro Industries, Manipur

The Thangjam Agro Industries, which is located at Chingmeirong, East Imphal, is a private sector. The unit started the processing of fruits from June, 1991. The unit produced and market their products under the brand name 'LIKLA' with FPO Licence No. 8327. The unit has 6 full time employees and 4 part time employees. Together 10 persons are employed. The main objective of this unit is to capture the market outside the state by taking the advantage of cheap and abundant availability of fruits in the state. Squashes of orange, pineapple and lemon are the main products of this unit.

4. Manipur Fruit Juice Crushers, Manipur

The Manipur Fruit Juice Crushers is located at M.G. Avenue, Imphal. The unit is under private sector and the same has started processing from 1981 under the brand name 'MANIFRU' with the FPO Licence No. 3564, the unit is manufacturing and marketing their products. The unit has altogether 18 persons employed, of which 8 are full time workers and 10 are part time workers. The main objective of this unit is to generate employment and economic upliftment of the state.

5. A Family of Food Products, Manipur

The brand name 'SANA' with FPO Licence No. 8321 is manufactured and marketed by M/S A Family of Food Products which is located at Sega Road, Takhellambam Leikai, Imphal. The unit is under private sector and has started processing of fruits from 10th October 1990. It gives employment opportunities to 11 persons altogether, of which 4 are full time workers and 7 workers are part time. The main objective of this unit is to solve the unemployment problems and to improve the economy of the state. Juices of pineapple, squashes of orange and pineapple, guava jelly and pineapple jam are the main products of this unit.

6. Deepashree Industries, Assam

The Deepashree Industries is located at D.K. Road, North Lakhimpur, Assam. This unit is under private sector and it has started processing of fruits from 1997. The unit is manufacturing and marketing their products under the brand name "DEEPA'S" with the FPO Licence No. 10320. The unit has 15 full time employees. The main objectives of the unit are to improve the economy of the state and to generate employment opportunities to the people. Pineapple juice, squash and jam are the main products of this unit.

7. Meghali Fruit Processing Unit, Assam

The Meghali Fruit Processing Unit is located at K.B. Road, North Lakhimpur, ward No. 4, Lakhimpur district, Assam. This unit is under private sector and had started processing of fruits from January 1983. The product of this unit is manufactured and marketed under the brand name 'MAJUDEE' with FPO Licence No. 7395. Six persons are employed by this unit. The main objectives of this unit are to process surplus fruits of local areas and to prevent wastage of fruits. Mango squash and litchi squash and also guava jam are the chief products of this unit.

8. The Hibi Fruit Products, Meghalaya

The Hibi Fruit Products manufactures and markets fruit products under the brand name "HIBI'S" with the FPO Licence No. 605. The unit is located at Krishna Mandir, Jail Road, Shillong, Meghalaya. This unit is under private sector and has started processing of fruits since 1955. The main objective of this processing unit is to generate income in order to support the family. The unit has 4 employees, of which 2 are full time workers and 2 are part time workers. Orange squash is the main item of this unit.

PART - II

The overall performance evaluation of fruit processing units can be broadly divided into four parts :

1. Data Analysis
2. Financial Analysis
3. SWOT Analysis
4. Hypothesis Testing.

Data Analysis : In order to study the performance of fruit processing units in different areas, a structured questionnaire was prepared which deals with financial management, marketing, production and general issues relating to fruit processing units of the region. The questionnaire was prepared and pre-tested in four units in Manipur. The questionnaire was finally administered to 54 fruit processing units located in Manipur, Assam, Arunachal Pradesh, Meghalaya, Mizoram, Nagaland and Tripura. These processing units deal with fruit juice, squash, jam, jelly, fruit R.T.S. etc. The units are being run under public, cooperative and private ownership. The data has been collected on the basis of first hand survey carried out in fruit processing units of the region. Effort was made to cover the entire fruit processing units in the region. Despite of the fervent effort made by the researcher, out of the total 54 units to whom questionnaire were administered, only 14 units could respond to the researcher's request. They are 8 units from Manipur, 3 units from Assam, 2 units from Meghalaya and 1 unit from Mizoram. Hence, an indepth analysis as to marketing management, production, finance and other general factors in developing fruit processing industry, success or failure has been carried out. The data analysis based on 14 units under study is as follows :

Form of Organisation

The below given Table-4.1 shows the form of organization wise profile of selected fruit processing units. Out of the selected 14 units, it is found that 8 units which accounted for more than 57 per cent of fruit processing units were under private sector.

Table 4.1

Form of Organisation-wise Profile of Fruit Processing Units

Organisation	*No. of Units*	*Percentage*
Private	8	57.14
Cooperative	3	21.43
Public	3	21.43
Total	14	100.00
	χ^{2*}	3.563

Source : Compiled on the basis of Questionnaire Administered to Fruit Processing Units.

* χ^2 represents Chi Square (see Appendix-V).

3 units each were under cooperative sector and public sector which accounts for 21.43 per cent of the sample units. The χ^2 was 3.563 which shows an insignificant variation between private sector, cooperative sector and public sectors.

Reasons for Location of the Units at the Present Site

To know about the reason for the choice of present location, a question was asked to the selected units. The data as per Table-4.2 indicates that fruit processing unit selected the present site because of the given reasons. Such as proximity of market centre, availability of raw materials, availability of infrastructural facilities and other factors.

It is observed from the below Table-4.2 that out of the total sample, 57 per cent of the units have selected the present site of location due to proximity to market centre, availability of raw materials and also infrastructural facilities. 15 per cent of the units have selected the present site because of availability of raw materials. The 7 per cent each have chosen the present site due to proximity to market centre and availability of raw materials, proximity to market only and availability of infrastructural facilities only. The remaining 7 per cent does not have response in this regard. The calculated value of χ^2 was 16.667 which is greater than the Table value of 11.1 for 5 degrees of freedom at 5 % level of significance which shows a significant variation in the above factors. Hence, we can conclude that proximity to market centre, availability of raw materials and availability of infrastructural facilities together were responsible for factors selecting the present fruit processing industry site.

Table 4.2

Reasons for the Choice of Present Industry Site

	Reasons	*No. of Response*	*Percentage*
A	Proximity to market centre	1	7
B	Availability of raw materials	2	15
C	Availability of Infrastructural facilities	1	7
D	All of the above	8	57
E	Only A and B	1	7
F	No response	1	7
	Total	14	100
		χ^{2*}	16.677

Source : Compiled on the basis of Questionnaire Administered to Fruit processing Units.

* χ^2 represents Chi Square (see Appendix-V).

Using Quality Control Technologies

Quality assurance is the key to every successful businessman even to Food Industries, quality control is the essence of a successful entrepreneurs. In order to know whether quality control technology were adopted in full measure by all the fruit processing units, a question was asked to the selected sample units. It was surprising to know that 28.57 per cent did not respond and the 7.14 per cent has outrightly responded 'No'. So only 64.29 per cent of the respondents were adopting quality control technology to improve the quality of their products. The value χ^2 was 6.986 which shows a significant variations between the users and non users of quality control technology in the region.

Table 4.3
Using Quality Control Technologies

Do you adopt Quality control technology to improve the quality of the product	*No. of response*	*Percentage*
Yes	9	64.29
No	1	7.14
No response	4	28.57
Total	14	100.00
	χ^{2*}	6.986

Source : *Compiled on the basis of Questionnaire Administered to Fruit Processing Units.*

* χ^2 represents Chi Square (see Appendix-V).

Production Management Problems

Efficient product management will eliminate the host of problems in any business. But the inefficient product management will lead to a less cost competitive in business operation. In order to know whether the fruit processing units of the North Eastern Region face any production problem or not, a question was asked to the selected sample units. It was found that 64.29 per cent of the respondents have production management problems.

The 21.43 per cent of the respondents have no production management problems. And the remaining 14.28 per cent have no response in this regard. The value of χ^2 was 6.129 which shows a significant difference in this regard.

Table - 4.4
Production Management Problems

Do you face any production problem	*No. of response*	*Percentage*
Yes	9	64.29
No	3	21.43
No response	2	14.28
Total	14	100.00
	χ^{2*}	6.129

Source : Compiled on the basis of Questionnaire Administered to Fruit Processing Units.

* χ^2 represents Chi Square (see Appendix-V).

Management of By-Product

Proper management of by-product will reduced the over all cost of production of the manufacturing units. It will definitely decrease the fixed cost of the manufacturing concern and hence the profit margin will increase. To know whether any by-product were manufactured by the fruit processing

Table - 4.5
Management of By-Product

Is there any by-product manufacture of your unit	*No. of response*	*Percentage*
Yes	2	14.29
No	12	85.71
No response	0	0
Total	14	100.00
	χ^{2*}	17.683

Source: Compiled on the basis of Questionnaire Administered to Fruit Processing Units.

* χ^2 represents Chi Square (see Appendix-V).

units, a question was asked to the selected sample units. It was found that only 14.29 per cent of the respondents have produced by-product from the waste materials after producing squash, jam and jelly. The remaining 85.71 per cent does not have any by product. The value of χ^2 was 17.683 which shows a significant difference in this regard.

Treatment of Waste Materials/Products

As it has been revealed by the above Table-4.5 that more than 85 per cent of the respondents does not have its by-product; significantly a question was also asked about the treatment of their waste products. It was found out that as many as 71.43 per cent of the respondents just throw away their waste products which other wise could be utilised for converting other by-product of the units. The 7.14 per cent have use it to feed animals and the remaining 21.43 per cent utilised the same for converting by-product like fruit candy. The χ^2 value was 9.555 which indicates a significant difference in this regard.

Table - 4.6
Treatment of Waste Materials/Products

How you treat your product waste	*No. of response*	*Percentage*
Throw away	10	71.43
Animal Feeds	1	7.14
Others	3	21.43
Total	14	100.00
	χ^{2*}	9.555

Source: Compiled on the basis of Questionnaire Administered to Fruit Processing Units.

* χ^2 represents Chi Square (see Appendix-V).

Conducting Market Surveys

Conducting Market Surveys are the foremost important responsibilities of a manufacturing unit in this modern world where customer is regarded as a king. Customers choices with regard to quality, quantity, size, etc. were determined in advance through market surveys. To know whether market surveys were conducted by the fruit processing units of the region before producing their product a question was asked to the selected sample units. It was found that only 28.57 per cent of respondents had a positive reply. Whereas 35.71 per cent had responded 'No' and another 35.71 per cent did not respond in this matter. The χ^2 value was just 0.139 which shows an insignificant difference in this regard.

Table - 4.7
Conducting Market Surveys

Have you ever conducted any market survey	*No. of response*	*Percentage*
Yes	4	28.58
No	5	35.71
No response	5	35.71
Total	14	100.00
	χ^{2*}	0.139

Source : Compiled on the basis of Questionnaire Administered to Fruit Processing Units.

* χ^2 represents Chi Square (see Appendix-V).

Channels of Distribution

Marketing function can be more effective with the adoption of appropriate channels of distribution. In order to know what are the different channels of distribution by the fruit processing units of the region a question was asked to the selected sample units.

Table - 4.8
Channels of Distribution

Sl. No.	*Channels of Distribution*	*No. of response*	*Percentage*
A.	Producer - Consumer	0	0
B.	Producer-Retailer-Consumer	3	21.43
C.	Producer-Wholesaler-Retailer-Consumer	8	57.14
D.	All of the above	2	14.29
E.	Only A and B	1	7.14
	Total	14	100.00
		χ^{2*}	11.057

Source : Compiled on the basis of Questionnaire Administered to Fruit Processing Units.

* χ^2 represents Chi Square (see Appendix-V).

It was found out that 57.14 per cent, the majority, of the respondents adopted the longest channel i.e. producer-wholesaler-retailer-consumer. Only 7.14 per cent of the respondents were using suggestible combined channels producer-consumer and producer-retailer-consumer. The 14.29 per cent of the respondents were adopting the given three channels together viz. producer-consumer, producer-retailer-consumer and producer-wholesaler retailer-consumer and the remaining 21.43 per cent of the respondents were adopting distribution channel of producer-retailer-consumer respectively. The value of χ^2 was 11.057 which shows a significant difference in this regard.

Fixation of Product Price

Product pricing is another important factor in marketing of consumer products. In order to know the methods of product pricing in fruit processing industry in North Eastern Region a structured question was asked to the selected sample units. It was found that 57.14 per cent of the respondents were still following the traditional method of product pricing viz. cost based pricing. The 14.29 per cent were adopting competition based pricing and the 21.43 per cent of the respondents were

Table - 4.9
Fixation of Product Price

Sl. No.	*Fixation of prices of product*	*No. of response*	*Percentage*
A.	Cost based pricing of product	8	57.14
B.	Demand based pricing	0	0
C.	Competition based pricing	2	14.29
D.	Only A and B	3	21.43
E.	Only B and C	1	7.14
	Total	14	100.00

Source : Compiled on the basis of Questionnaire Administered to Fruit Processing Units.

adopting a combination of cost based pricing and competition based pricing. The remaining 7.14 per cent of the respondents were adopting a combination of demand based pricing and competition based pricing as the method of pricing their products.

Selling Techniques adopted by the Fruit Processing Industry

Selling techniques are indispensable with the successful marketing of consumer products including processed fruit products. In order to know the selling techniques adopted by the fruit processing units of the North Eastern Region, a question was asked to the selected sample units. It was found that 50 per cent of the respondents were adopting the given three techniques together viz. personal, advertising through different medias and sales promotion.

Table - 4.10
Selling Techniques adopted by the Fruit Processing Units

Sl. No.	*Selling Techniques*	*No. of response*	*Percentage*
A.	Personal	3	21.43
B.	Advertising through different medias	0	0
C.	Sales promotion	1	7.14
D.	All of the above	7	50.00
E.	Only A and B	2	14.29
F.	Only A and C	1	7.14
	Total	14	100.00
		χ^{2*}	11.031

Source : Compiled on the basis of questionnaire administered to fruit processing units.

* χ^2 represents Chi Square (see Appendix-V).

The 21.43 per cent of the respondents were adopting personal contact selling techniques. The 14.29 per cent of the respondents were adopting combined techniques of personal contact and advertising through different medias and the remaining 7.14 per cent each were adopting sales promotion and a combination of personal selling and sales promotion techniques. The χ^2 value was 11.031 which shows an insignificant difference in adopting selling techniques among the fruit processing units.

Packaging Strategy Adopted by the Fruit Processing Industry

Packaging is one of the important components of successful marketing. To know the prominent strategy of fruit processing unit of the North Eastern Region, a question was asked to the selected sample units. It was found that 35.71 per cent of the respondents are still following the traditional

Table - 4.11
Packaging Strategy adopted by the Fruit Processing Unit.

Sl. No.	*Packaging Strategy*	*No. of response*	*Percentage*
A.	Plastic bottle	3	21.43
B.	Glass bottle	5	35.71
C.	Can	0	0
D.	All of the above	3	21.43
E.	Only B and C	3	21.43
	Total	14	100.00
		χ^{2*}	4.96

Source : Compiled on the Basis of Questionnaire Administered to Fruit Processing Units.

* χ^2 represents Chi Square (see Appendix-V).

packaging strategy viz. the glass bottle. The 21.43 per cent each of the respondents were adopting only plastic bottle, combination of plastic bottles, glass bottles and can and a combination of glass bottle and can only. None of the units were adopting the modern packaging strategy like asceptive packaging. The χ^2 value was 4.96 which shows an insignificant variations in this regard.

Participation in Trade Fairs

Modern way of promoting the product specially the new product is through participating in trade fairs. In order to know whether the fruit processing units of the region participated in trade fairs or not a question was asked to the selected sample units. It was found out that 92.86 per cent of the respondents have already participated in the trade fairs where as 7.14 per cent have not participated in any trade fairs. The χ^2 value was 17.742 which shows a significant difference in this regard.

Table - 4.12
Participation in Trade Fairs

Have you participated in any trade fair	*No. of response*	*Percentage*
Yes	13	92.86
No	1	7.14
No response	0	0
Total	14	100.00
	χ^{2*}	17.742

Source : Compiled on the basis of Questionnaire Administered to Fruit Processing Units.

* χ^2 represents Chi Square (see Appendix-V).

Sources of Financing

Finance is the life and blood to business. So to run the industry, finance is required. In order to know about the sources of financing fruit processing industry in North Eastern Region, a question was asked to the selected sample units. Based on this questionnaire, sources of financing can be classified into three groups. The first group were the government and agencies like Ministry of Food Processing Industries, Govt. of India, National Cooperative Development Corporation (NCDC), EZE Germany (Foreign Agency) and other state agencies. Banking and financial institutions like State Bank of India (SBI), Central Bank of India, Small Industrial Development Bank of India (SIDBI), Manipur State Cooperative Bank were the second group. Third and last group were family and own source. From the Table-4.13, it was found out that 28.57 per cent each of the respondents were family and from their own sources and a combination of banking institutions and government and agencies respectively. The 21.43 per cent of the respondents were funded by bank and financial institutions. The 14.29 per cent of the respondents were financed by the Government and

Agencies. And the remaining 7.14 per cent of the respondents were together financed by banking institutions, government agencies and from own sources. The χ^2 value was 2.557 which shows an insignificant difference in financing of this particular sector of business.

Table - 4.13
Sources of Financing

Sl. No.	*Sources of financing*	*No. of response*	*Percentage*
A.	Bank and Financial Institutions	3	21.43
B.	Government and Agencies	2	14.29
C.	Family and from own sources	4	28.57
D.	All of the above	1	7.14
E.	Only A and B	4	28.57
	Total	14	100.00
		χ^{2*}	2.557

Source : Compiled on the basis of Questionnaire Administered to Fruit Processing Units.

* χ^2 represents Chi Square (see Appendix-V).

Availability of Loan Facilities from the Banks and Financial Institutions

To know whether Bank Loans were available to the fruit processing units of the region, a question was asked to the selected sample units. It was found out that 50 per cent of the respondents were not getting loan facilities from banks and financial institutions. The 42.86 per cent said that loan facilities were made available to them. The χ^2 value was 4.417 which shows an insignificant variations as to availability and non availability of loan facilities to the fruit processing units.

Table - 4.14
Availability of Loan Facilities from Banks and Financial Institutions

Was the loan made available to you	*No. of response*	*Percentage*
Yes	6	42.86
No	7	50.00
No. response	1	7.14
Total	14	100.00
	χ^{2*}	4.417

Source : Compiled on the basis of Questionnaire Administered to Fruit Processing Units.

* χ^2 represents Chi Square (see Appendix-V).

Government's Attitude Towards Fruit Processing Units

If we are to develop fruit processing industry to the fullest measure, particularly in the backward region like the North Eastern Region, the role of both the state and the central

Table - 4.15
Government's Attitude Towards Fruit Processing Units

Government's Role	*No. of response*	*Percentage*
Very Helpful	1	7.14
Helpful	3	21.43
Indifferent	4	28.57
Not Helpful	2	14.29
No Response	4	28.57
Total	14	100.00
	χ^{2*}	2.428

Source : Compiled on the basis of Questionnaire Administered to Fruit Processing Units.

* χ^2 represents Chi Square (see Appendix-V).

government for its development is inevitable. In order to know the government's attitudes towards fruit processing units of the North Eastern Region, a question was asked to the selected sample units. It was found out that only 7.14 per cent of the respondents says that government are very helpful to their units. The 21.43 per cent of the respondents responded that government are helpful to their units. But the 14.29 per cent of the responded outrightly pointed out that government are not helpful to their units.

The 28.57 per cent of the respondents have no response in this regard. The χ^2 value was 2.428 which shows an insignificant variation in this regard.

Future Strategy to Meet the Challenges

In the wake of WTO Agreement, success of the business specially manufacturing concern like fruit processing units depend mainly on their strategy. In order to know the future

Table - 4.16

Future Strategy to Meet the Challenges

Future Strategy	*No. of response*	*Percentage*
Combined with the growers	4	28.57
Combined with the other fruit processing unit	4	28.57
Combine with other food manufacturing unit	1	7.14
No. response	5	35.72
Total	14	100.00
	χ^{2*}	2.577

Source : Compiled on the basis of Questionnaire Administered to Fruit Processing Units.

* χ^2 represents Chi Square (see Appendix-V).

strategy to meet the challenges of fruit processing industry of the North Eastern Region, a question was asked to the selected sample units. It was found out that 28.57 per cent each of the respondents have the future strategy of combining with the growers and combining with the other fruit processing unit in order to meet the challenges arising after the implementation of WTO Agreement in the country. The 7.14 per cent of the respondents have the future strategy of combining with other food manufacturing unit. But the remaining 35. 72 per cent has no response in this regard. The χ^2 value was 2.577, which shows an insignificant difference with regard to future strategy.

To sum up, the form of organisation with regard to fruit processing units in the North Eastern region indicates that majority of the units are under private sector. Proximity to market centre, availability of raw materials and availability of infrastructural facilities play an important role in determining the site of fruit processing industry. The overall data analysis indicates that the fruit processing units of the north eastern region faces mainly the three facets problems such as financial management, production and marketing. The first hand survey points out that majority of the fruit processing units of the region does not conduct market survey to know the need and demand of the customer.

PART - III

Financial Analysis : To analyse the financial performance of the fruit processing units, ratio analysis was taken as it is the most important and powerful tool. A ratio is defined as 'the indicated quotient of two mathematical expressions and as the relationship between two or more things'[1]. Absolute accounting information in the annual reports do not provide much information about the financial matters. A meaningful conclusion can be drawn only when the relationship between the two figures are calculated i.e. ratios. In this part, ratio analysis is used as an index or yard stick for evaluating the financial performance of the fruit processing units under study.

Further, among the different sub-categories of ratio analysis, profitability measures analysis has been selected for the present study. Under this measure again four ratio were used such as Gross Profit Ratio, Net Profit Ratio, Return on Assets and Return on Capital Employed.

Besides, Owners and management of the fruit processing units, creditors and financial institution want to know the financial soundness especially the profitability of the firm. Creditors and financial institutions want to get interest and repayment of principal regularly. On the other hand, owners and proprietor invest their resources with an expectation of higher rate of return on their investment. In addition to this, the management of the units are eager to know the managerial efficiency of the firm. The motives behind all these parties ultimately depends on profit of the units. So profit plays the important yardstick for success or failure over a period of time.

Profitability of a firm can be measured by profitability ratios. It can be determined on the basis of either sales or investment. The profitability ratios in relation to sales are (i) Gross Profit Ratio (ii) Net Profit Ratio. The profitability in relation to investment is measured by (i) Return on Assets (ii) Return on Capital Employed.

Profitability Ratios Related to Sales : These ratios are based on the principle that a firm should earn adequate profit on each rupee of sales. If the fruit processing unit is unable to earn sufficient profit on sales, there will be difficulty in meeting the operating expenses, interest charges and as a result no return will be available to owners. As already mentioned these ratios consists of i) Gross profit ratio and ii) Net profit ratio.

i) Gross Profit Ratio (GPR): The first profitability ratio in relation to sales is the gross profit ratio. It is calculated by dividing the gross profit by sales.

$$\text{Gross Profit Ratio} = \frac{\text{Gross Profit}}{\text{Sales}}$$

The Gross Profit Ratios of 14 fruit processing units has been shown in Table-4.17. The brand name "DEEPA'S" of Deepashree Industries, Lakhimpur district of Assam stood the first rank in GPR with 33.33 per cent. Followed by the brand name 'MAGFRUIT' of Govt. Fruit Preservation Factory, Imphal, Manipur and 'SANA' of A Family of Food Products, Imphal, Manipur having the percentage of 25.93 and 22.86 respectively. The Gross Profit Ratio of the brand name 'HEIRANG' of the Manipur Fruit Processing & Cold Storage Co-operative Society Ltd., Imphal Manipur is the least with just 6.9 per cent.

ii) Net Profit Ratio (NPR) : The Net Profit Ratio is calculated by dividing net profit by sales :

$$\text{Net Profit Ratio} = \frac{\text{Net Profit}}{\text{Sales}}$$

The Table-4.17 indicates that the Net Profit Ratio of the brand name "DEEPA'S" of Deepashree Industries, Lakhimpur district of Assam again stood the first rank with 5 per cent. It is followed by the NPR of the brand name 'MANIFRU' of Manipur Fruit Juice Crushers, Imphal, Manipur and the brand name 'MEG' of Govt. Fruit Preservation Factory, Shillong, Meghalaya having the percentages of 3.36 and 2.55 respectively. The NPR of the brand 'RATNA' of Ratna Fruit India, Imphal, Manipur is the least with just 0.19 per cent.

In comparing the two ratios, the profitability ratios of the brand "DEEPA'S" of Deepashree Industries, Lakhimpur district of Assam continues to topped the ranking. But the GPR and NPR continues to fluctuate for the other brand. For instance, the brand 'HEIRANG' of Manipur Fruit Processing and Cold Storage Co-operative Society Ltd. Manipur is the least in GPR but stood 9th position in the NPR.

Profitability Ratios related to Investment : Profitability of the fruit processing units can also be calculated based on the profits of the units to its investments. Such ratios are known as return on investments. Profitability in relation to investment is

Table - 4.17

Rank Analysis of Profitability Ratios of Selected Fruit Processing Units of N.E. India

Sl. No.	Brand Name of the Units	G.P. Ratio		N.P Ratio		ROA Ratio		ROCE Ratio		Overall Performance (in rank)
		Average	*Rank*	*Average*	*Rank*	*Average*	*Rank*	*Average*	*Rank*	
1.	Chandni	10.00	10	0.30	13	0.52	9	0.75	9	12
2.	Deepa's	33.33	1	5.00	1	9.75	1	12.19	1	1
3.	Heirang	6.90	14	0.62	9	1.08	6	1.23	6	8
4.	Hibi	9.09	11	0.45	12	0.75	7	0.83	8	10
5.	Likla	16.97	6	1.04	6	0.10	13	0.11	13	9
6.	Magfruit	25.93	2	1.37	4	2.56	4	3.29	4	4
7.	Majudee	11.43	9	1.00	7	1.14	5	1.79	5	5
8.	Manifru	20.00	5	3.36	2	7.83	2	10.53	2	2
9.	Meg	21.21	4	2.55	3	3.77	3	4.58	3	3
10.	Mifco	16.67	7	1.12	5	0.43	10	0.56	10	6
11.	Niru's	15.00	8	0.63	8	0.04	14	0.05	14	11
12.	Ratna	7.45	12	0.19	14	0.16	11	0.18	11	13
13.	Sana	22.86	3	0.49	11	0.67	8	0.82	7	7
14.	Waifruit	7.41	13	0.59	10	0.11	12	0.13	12	14

Sources: Compiled on the basis of - 1. Financial statements given by the units
2. Questionaire Administered to Fruit Processing Units

measure by (i) Return on Assets and (ii) Return on Capital Employed.

i) Return on Assets (ROA): The relationship between profit and assets can be shown by the return on assets ratio. This ratio can be calculated by dividing net profit by total assets.

$$\text{ROA} = \frac{\text{Net Profit}}{\text{Total Assets}}$$

Here net profit is the profit after tax and total assets include net fixed and current assets.

In the above Table-4.17, the ROA of the brand "DEEPA'S" of Deepashree Industries, Lakhimpur district of Assam stood the first rank with 9.75 per cent. The brand 'MANIFRU' of Manipur Fruit Juice Crushers, Imphal and 'MEG' of Govt. Fruit Preservation Factory, Meghalaya stood the second and third rank in ROA ratio with 7.83 per cent and 3.77 per cent respectively. The ROA ratio of "NIRU'S" of Guwahati is the least with 0.04 per cent.

ii) Return on Capital Employed (ROCE) : The second type of ROI is the ROCE. This ratio has similarities with ROA except in one respect i.e., here the profits are related to the total capital employed. The term capital employed refers to long term fund supplied by the creditors and owners of the firm. In other words, it is equivalent to net working capital plus fixed assets.

Thus, ROCE can be computed in the following manner.

$$\text{ROCE} = \frac{\text{Net Profit after tax}}{\text{Total capital employed}}$$

From the above Table-4.17, the ROCE of the brand "DEEPA'S" of Deepashree Industries, Lakhimpur district of Assam continues to topped the ranking with 12.19 per cent. It is followed by the brand 'MANIFRU' of Manipur Fruit Juice Crushers, Imphal and 'MEG' of Govt. Fruit Preservation Factory, Meghalaya ranked 2nd and 3rd with 10.53 per cent and 4.58 per cent respectively. The ROCE of "NIRU's" of Modern Fruit Processing Co-operative Society Ltd., Guwahati continues to be the least in ranking.

By comparing the four parameters namely, GPR, NPR, ROA and ROCE, we can conclude that the brand "DEEPA'S" of Deepashree Industries Lakhimpur district of Assam can be ranked first followed by the brand 'MANIFRU' of Manipur Fruit Juice Crushers and the brand 'MEG' of Govt. Fruit Preservation Factory, Meghalaya having the ranking of second and third respectively. The brand 'WAIFRUIT' of Agrotech and Aqua Products, Thoubal district of Manipur stood the least in the overall ranking of the profitability ratios of the fruit processing units in the North Eastern Region.

PART - IV

SWOT Analysis of Fruit Processing Units

SWOT or WOTS is an acronym for strengths, weaknesses, opportunities and threats. This analysis is the combination of organisational analysis and environmental analysis which is undertaken as an integrated process in strategic management. The basic objective of SWOT analysis is to provide a frame work to reflect on the organisational capabilities to avail opportunities or to overcome threats presented by the environment. In fact, the dimensions of organizational capabilities have relevance in so far as they relate to the environmental conditions. Hence, organizational analysis and environmental analysis should be interlinked.

SWOT analysis presents the information about external and internal environment in structured form whereby key external opportunities and threats can be compared systematically with internal strength and weakness. Thus, organisation's external and internal situations can be matched so as to form distinct patterns and thus, strategy can be chosen on the basis of these patterns. These patterns may be in the form of combinations of the following :

i) High opportunities and high strengths

ii) High opportunities and low strengths

iii) High threats and high strengths

iv) High threats and low strengths.

Naturally, in all these situations, different strategies are suitable, for example, in the case of first situation, aggressive growth strategy may be suitable, while defensive strategy like divestment strategy may be suitable in fourth situation.

Fruit processing units in North Eastern Region has shown a high opportunities due to favourable agro-climatic conditions to grow different types of fruits and the technology offer in this modern era, can surely compete with others particularly in this sub-sector. But due to lack of adequate working capitals, lack of intelligence, lack of infrastructural facilities and the like has revealed a low strength in this sub-sector. In such situation, it will be apt to apply the economic strategy where resources and opportunities are matched in a proper way so as the internal strength can be improved.

The agragrian base of India holds great promise with regard to processing of fruits and this segment of food processing industry is expected to grow fairly in coming years. The fruits because of their high productivity and value offer much higher economic returns per hectare acreage compared to cereal crops. They also have good export potential. The prominent items processed in the fruit processing sector are fruit pulps, fruit based ready to serve beverages, canned fruits, jams, juices, squashes, pickles etc. Major processing unit of North East India's characteristics relevant to identification of opportunities, threats, strengths and weaknesses are as follows:

Opportunities : The following are the basic opportunities avail in Fruit Processing Units of the North Eastern Region:

1. Processed fruit products of the region have a growing market demand specially in urban area of the region in particular and India in general.

2. The North Eastern Region is bordering with the four countries of China, Bhutan, Myanmar and Bangladesh and this offers a unique opportunities for export of processed fruits products to these countries.
3. This sub-sector requires lower investment, from the individual and state levels, if compared with the other capital intensive industrial sector.
4. The production of fresh fruits can be increased manifold if the farm practices of the region are improved specially in the field of pre and post harvest activities of fruits plantation.
5. India's relationship with the neighboring countries will provide a boost to trade and industrial activity of the region.
6. Development of modern technology offers greater opportunity to compete with the products of national market and international market.

Threats : The following are the pertinent threats to fruit processing units of the North Eastern Region:

1. Implementation of WTO Agreement will give greater threat to wipe out small and inferior manufacturer or processor from the export and domestic market.
2. Natural factors like the terrain and floods can disrupt the industry completely.
3. Law and order situation of the insurgency prone region which is in a pathetic condition, can easily cripple the industrial performance if it continue to do so.
4. With the development of horticultural techniques, the region may lag behind the other states in the country which become more competitive in production and quality.

Strengths : The internal strength of the fruit processing units of the North Eastern Region are as follows:

1. The region is extremely suitable for horticultural production, because of the agro-climatic conditions and the terrain.
2. Fruits of different varieties are available at low prices and in high volumes during the season.
3. The raw material supply situation is favourable since the location of the processing units is in the fruit growing belt.

Weaknesses : The internal weaknesses of the fruit processing units of the region are given below:

1. Infrastructure deficiencies, in terms of transport, communication, power and post harvest facilities, are high.
2. Existence of many middlemen in the supply side of the chain, leading to reduce profit margins of the growers in one side and fruit processor on the other side.
3. High cost of packaging and transportation can increase the cost of production per unit of processed fruits.
4. Brands are not well established and have a local presence only.

To conclude, the overall scenario emerging after the analysis is that the fruit processing sector in the region presents attractive opportunity for investment despite the poor performance of the existing units. There are, if not many, few units performing well in this sector and these units should further improve the quality of their product so that the products could be exported even to international markets. However, this must be preceded by proper planning and a focussed approach towards the market requirements, especially for an export oriented unit in the region.

PART V

Test of Hypothesis

To test the validity of assumption, it is required to gather sample data and determine the difference between the hypothesized value and the actual value of the sample mean. The smaller the difference, the greater the likelihood that our hypothesized value for mean is correct. The larger the difference, the greater the likelihood that our hypothesized value for mean is incorrect.

The null hypothesis is a very useful tool in testing the significance of difference. In its simplest form the hypothesis asserts that there is no real difference in the sample and the population in the particular matter under consideration (hence the word "null" which means invalid, void or amounting to nothing) and that difference found is accidental and unimportant arising out of fluctuations of sampling.

The rejection of the null hypothesis indicate that the differences have statistical significance and the acceptance of the null hypothesis indicates that the differences are due to chance. Since many practical problems aim at establishment of statistical significance of differences, rejection of the null hypothesis may thus indicate success in statistical project. In statistical terms a null hypothesis may be stated as :

$$Ho : \mu_1 = \mu_2$$

Where μ_1 and μ_2 are population means.

It states that there is no significant difference in the means of the two populations. If facts lead to the rejection or non retention of the Ho, then the alternative hypothesis H_1 as stated below stands accepted.

$H_1 : \mu_1 \neq \mu_2$ in which $\neq$ means "not equal to" and μ_1 and μ_2 are population means. It means that the two populations differ significantly.

The statistical analysis for correlation coefficient is calculated directly and difference test/paired T- Test is calculated by using computer software package i.e. Statistical Package for Social Science (SPSS).

To test the hypothesis, whether increase in transportation cost leads to increase in cost of production per unit of fruit processing industry, Karl Pearson's Correlation Coefficient equation was applied on the collected data of 8 fruit processing units from Manipur which is given in Table-4.18.

Table-4.18
Cost Price Per Unit & Transportation Cost Per Unit

Name of the Brand	*Cost price per unit*	*Transportation cost per unit*
Chandni	27	1.38
Heirang	27	1.40
Likla	20	1.07
Magfruit	31	1.60
Manifru	28	1.42
Ratna	25	1.18
Sana	27	1.39
Waifruit	25	1.16

Source : Based on Questionnaire collected from Fruit Processing Units.

Table-4.18.1
Calculation of Karl Pearson's Correlation Coefficient

X	Y	X^2	Y^2	XY
27	1.38	729	1.904	37.26
27	1.40	729	1.960	37.80
20	1.07	400	1.145	21.40
31	1.60	961	2.560	49.60
28	1.42	784	2.016	39.76
25	1.18	625	1.392	29.50
27	1.39	729	1.932	37.53
25	1.16	625	1.345	29.00
210	10.6	$\Sigma x^2 = 5582$	$\Sigma y^2 = 14.225$	$\Sigma XY = 281.85$

$n = 8$

$\overline{X} = 26.25 \ \overline{Y} = 1.325$

$$r = \frac{\Sigma XY - n\overline{X}\overline{Y}}{\sqrt{\Sigma X^2 - n\overline{X}^2}\sqrt{\Sigma Y^2 - n\overline{Y}^2}}$$

Now putting the value into the equation, we get

$$r = \frac{281.85 - 8 \times 26.25 \times 1.325}{\sqrt{5582 - 8 \times (26.25)^2}\ \sqrt{14.225 - 8 \times (1.325)^2}}$$

$$= \frac{281.85 - 278.25}{\sqrt{5582 - 5512.5}\ \sqrt{14.225 - 14.045}}$$

$$= \frac{3.6}{\sqrt{69.5}\ \sqrt{0.21}}$$

$$= \frac{3.6}{8.337 \times .458}$$

$$= \frac{3.6}{3.818} \quad = 0.942$$

From the above correlation analysis, it has been found that the cost of production per unit of fruit processing units is positively associated with transportation cost per unit. The calculated value of r is 0.942 which is highly significant at 1% level of significance.

Therefore, increase in transportation cost per unit will lead to increase the cost of production per unit, hence the hypothesis is true.

To test whether fruit processing units were born sick because of lack of infrastructural facilities or not, the major infrastrutural facilities such as power, banking facilities, transportation and communication of the North Eastern Region were discussed in detail in Table-2.3 of Chapter 2. And it has already been found that the infrastrutural facilities in North

Eastern Region are far behind than all India average. To test whether it is significant or not, one of the important infrastruture i.e. power has been taken. In this connection 8 units out of 14 units were selected and a pertinent question was asked to each of the fruit processing unit about the power required and power available. Based on the data collected from these 8 units paired T test was done through computer software package (SPSS) and the following results were drawn :

Table - 4.19

Power Required and Power Available

(Power in HP)

Name of the Brand	*Power Required*	*Power Available*
CHANDNI	3	1.875
HEIRANG	3	3
LIKLA	3	2
MAGRUIT	3	2.5
MANIFRU	8	5
RATNA	3	2
SANA	3	1.875
WAIFRUIT	3	2

Source : Based on Questionnaire collected from Fruit Processing Units.

Table - 4.19.1

Paired T-Test

(Power Required & Power Available)

Paired Samples Statistics

		Mean	*N*	*Std. Deviation*	*Std. Error Mean*
Pair	P_REQD	3.6250	8	1.7678	0.6250
1	P_AVLE	2.5313	8	1.0706	0.3785

Note : P_REQD = Power Required; P_AVLE = Power Available; Std. = Standard

Table - 4.19.2
Paired Samples Correlations

	N	*Correlation*	*Sig.*
Pair 1 P_REQD & P_AVLE	8	0.932	0.001

Note : P_REQD = Power Required; P_AVLE = Power Available; Sig. = Significance

Table - 4.19.3
Paired Samples Test

	Paired Differences							
				95% Confidence Interval of the Difference				
	Mean	*Std. Deviation*	*Std. Error Mean*	*Lower*	*Upper*	*t*	*df.*	*Sig. (2-tailed)*
Pair 1 P_REQD-P_AVLE	1.0938	0.8628	0.3050	0.3724	1.8151	3.586	7	0.009

Note : P_REQD = Power Required; P_AVLE = Power Available; Std. = Standard;

t = Calculated Value; df. = Degree of Freedom; Sig. = Significance

From the above result obtained from the computer analysis, it is found that the daily mean power required per unit and mean power available per unit are 3.625 and 2.5313 HP respectively and standard deviation of power required per unit and power available per unit are 1.7678 and 1.0706 respectively.

Paired sample correlation between the two variable has been found to be 0.932 which is highly significant.

The paired sample Test for the diffcrence in mean of per unit power required and per unit power available has been found to be significant at 1 % level.

Therefore, the present power available to fruit processing units of the region are not upto the level of expectation, hence the hypothesis is true.

To test whether outside NE region the brand of processed fruits captured more than 50 per cent of the total market of fruit squash, juice, jam and jelly than the local brand producing within the region 56 retailers having their shops in different market places both Imphal East and Impal West district were interviewed which is presented in Appendix - VIII. Based on the response given by 56 retailers, following result were derived through computer packages of SPSS.

Table - 4. 20.1
Paired T-Test
(Local & Outside Brand)
Paired Samples Statistics

		Mean	*N*	*Std. Deviation*	*Std. Error Mean*
Pair	Local	39.5000	56	38.9134	5.2000
1	Outside	59.1250	56	114.7890	15.3393

Note : Std. = Standard

Table - 4. 20.2
Paired Samples Correlations

	N	*Correlation*	*Significance*
Pair 1 Local & Outside	56	0.261	0.052

From the above result of analysis it is found there is a great deal of difference in the mean of local and outside brand . This has been supported by a paired sample correlation of 0.261, which is significance at 1 % level. However, a statistical test for paired difference of mean of sale between local and outside brand has not found to be significant. The main reason behind it was due to seasonal demand of processed fruits.

Table - 4.19.3
Paired Samples Test

	Paired Differences							
				95% Confidence Interval of the Difference				
	Mean	*Std. Deviation*	*Std. Error Mean*	*Lower*	*Upper*	*t*	*df.*	*Sig. (2-tailed)*
Pair 1 Local-Outside	**19.6250**	**111.1635**	**14.8549**	**49.3948**	**101448**	**-1.321**	**55**	**0.192**

Note : Std. = Standard; t = Calculated Value; df. = Degree of Freedom; Sig. = Significance

REFERENCE

1. *Webster's New Collegiate Dictionary*, 8th Edition Springfield, Mass G&C Merrain, 1975, p.958.

5

Finding of the Study and Suggestions

Introduction

The form of organisation with regard to fruit processing units in the North Eastern Region indicates that majority of the units are under private sector. In the entire region, 68.85 per cent are under the private sector form of organisation. This shows that private sector plays the important role in the development of fruit processing industry in the region.

The overall performance evaluation reveals that the fruit processing units of the North Eastern Region faces mainly the three facets problems such as financial management, production and marketing. Shortage of working capital for procurement of raw materials and packaging materials which are not available in small units were the main problem of financial management. None availability of sufficient raw materials of good quality and adequate power supply were the main problem in producing a standard quality. Poor sales and marketing performance were experienced by the units of this region. The first hand survey reveals that majority of the fruit processing units of this region does not conduct market survey to know the needs and requirements of the customers.

A new orientation needs to be given to the fruit processing industry to upgrade quality, minimise post harvest losses, ensure remunerative returns to the growers as well as to the processors, increase employment avenues in the entire North Eastern Region and generate foreign exchange earnings.

PART - I

Findings of the Study

The following are the main findings of the present study:

1. Private sector plays an important role in the development of fruit processing industry in North Eastern States. Assam, Manipur, Meghalaya and Nagaland have 88, 69.2, 75 and 60 per cent of private sector fruit processing industry respectively. In the entire region 68.85 per cent are under the private sector form of organisation.

2. The main sources of financing fruit processing industry in North Eastern Region can be broadly classified into three groups. The first group were the government and other agencies like MFPI, Govt. of India, National Cooperative Development Corporation (NCDC), EZE Germany (Foreign Agency) and other state agencies. Banking and financial institutions like State Bank of India, Central Bank of India, Small Industrial Development Bank of India (SIDBI), Manipur Cooperative Bank were the second group. Third and last group were family and own source. It was found out that family and own sources contributes the highest with 28.57 per cent share in the financing of fruit processing industry in the region.

3. The overall performance ranking of fruit processing units in the North Eastern Region indicates that the brand 'DEEPA'S' of Deepashree Industries, Lakhimpur district of Assam stood first. It is followed by the brand 'MANIFRU' of Manipur Fruit Juice Crushers, Imphal and the brand 'MEG' of the Govt.

Fruit Preservation Factory, Meghalaya occupying the second and third position respectively. The brand 'WAIFRUIT' of Agrotech and Aqua Products, Thoubal district of Manipur stood the least in the overall ranking of the profitability of ratios of fruit processing units in the North Eastern Region.

4. Regional branding is absent, with each state and each unit going for its own branding. This dilutes the promotion effort and forces the brands into becoming local players only.
5. Post harvest facilities from the fruit plucking stage to the processing practices and cold stores facilities are very much lacking. It has been told that the farmers shake the tree in order to pluck the fruit from the tree. This thumb rule method has not only increased the wastage but also reduced the quality of the raw material.
6. Non availability of suitable raw materials in terms of size, colour, texture etc., lack of awareness about the processable varieties, low yield and poor quality of raw materials are the main causes which leads to low productivity of the units.
7. Availability of packing materials is erratic and the procurement is done from the Kolkatta and Guwahati market. The cost of the packaging material is also very high.
8. Marketing efforts are not well organised to counter the competitor from well established national brands, such as Kissan, Bikaji and Rallies and products of third world countries importing through Moreh.
9. Lack of financial planning leading to short sighted decisions related to financial assistance, loans and subsidies. The financial measures are generally aimed at solving immediate problems of the units rather than the basic causes of those problems.

10. The technology is outdated as most of the plants had been commissioned many years back. This affects their efficiency and raises the costs of maintenance and repair.
11. The scale of operation is very small, this can neither absorb a significant proportion of the horticulture produce nor can it achieve the economies of scale for a national level market entry.
12. The power supply, transport and other infrastructural facilities of the region in general and particularly Manipur is inadequate to meet the requirement of these units.
13. Inadequacy of working capital leads to constraints with the sound financial management which affects the day to day workings of the units.
14. Trends of the Indian food market were not easily adopted by these units because of the isolation of the region from the mainland, the local presence of the regional products and the financial constraints of these units.
15. There is lack of awareness and information on WTO provisions among the processing units of the region. Hence, they hardly prepare to tackle the situation in the event of WTO implementation particularly in India being a member country.
16. There is an absence of stable linkage between the grower and the processing unit and also between the market and the processing units.
17. The processing units of this region has little knowledge about the national market and international market and therefore, at present, they are not trying to enter either national or international market.

18. Most of the units are still following the traditional method of marketing and so they are not fulfilling the customer's desire and needs.

19. Lack of adequate subsidy, grants and schemes to the poor farmers has resulted in less development of household industries in the region.

20. Though various factors including insurgency, dearth of raw materials and skilled workers, communication bottlenecks etc. have been hold responsible for generating a depression prone atmosphere, inadequate supply of power is considered to be a major deterrent to the growth of the fruit processing industry in the region.

21. The role of co-operative sector is negligible as its contribution to this sub-sector is just 8 per cent in the region. This could be one of the reason for slow progress of processing units in the region where there is vast potentiality in growing horticulture crops particularly the fruit crops.

22. The attention of the Indian corporate sector focuses on emerging opportunities in the North East. This has been identified as agro and food processing sectors which includes fruit processing industry in particular and horticulture in general, can usher in a green revolution as the region offers diverse agro-climatic conditions for growing a variety of crops which have rich export potential.

23. India has one of the highest tax rate charged against the food processing industries as compare to other Asian Countries. It is the only country to levy excise duty on processed foods apart from Thailand.

24. Post harvest technology, infrastructure building, cold chains and packaging are some of the thrust areas in which India is seeking competent overseas collaboration for food processing industries.

25. The estimated production of fruits for 2002-03 in the country is 33 million tonnes which account only 8 per cent of total world production of fruits next to China. Calculating 30 per cent post harvest losses on the above estimate, the total losses of India will be 9.9 million tonnes. Where as processing of 2 per cent of fruits in the above estimate will be 0.66 million tonnes. Therefore, the remaining 22.44 million tonnes of fruits are supposed to have marketed and consumed in fresh form.

PART - II

Suggestions

Despite all difficulties and constraints discussed in the above, the scope for expansion in fruit sector is enormous. Some of the suggestions on the basis of the above study are given as under :

1. More fruit processing units should be set up in North Eastern Region. These units could be tied up with large and established companies like Hindustan Lever Ltd. and Dabur which has been successfully marketing fast moving consumer goods like fruit products in India and abroad for the past many years so as to get wider market for the products of this region.

2. Cooperative marketing which is considered as the 'golden rule' in agriculture marketing should be fully developed in the region for both the fresh fruits and the processed fruit products.

3. The existing fruits processing units in North East India should try to create a uniform product line so that the marketing can be done under a single brand name. The units need to practice uniform production practices so that the quality will be standardised and the marketing agency should have enough supply pool to gather.

4. There is a need to increase manufacture of safe, hygienic, nutritive and better quality value added processed products conforming to global standards at competitive and economic cost.

5. Training centres should be created for the farmers in the area of post harvest management of agro produce to encourage creation of preprocessing facilities close to the area of production to avoid wastage and to reduce cost of transportation of raw material. Processing units, can collect products from the doorsteps of growers. Alternatively farmers can take it to the industry before it perishes.

6. It is most essential to optimally utilise waste of processing units for manufacture of food items, animal feeds, fertilizer (green manure) as well as other industrial products. This will not only improve the economies of operation but will also be environment friendly.

7. North Eastern Region of India is bordering with four other nations namely, China, Myanmar, Bangladesh and Bhutan and this offers the unique opportunities and potential for export trade not only to these countries but also to other South East Asian Countries. Therefore, policies should be formulated in such a way that the North Eastern States should be benefited out of this.

8. Government level agreement with Bangladesh, regarding road/rail transport through Bangladesh to Kolkatta may once get finalised. This will provide a boost to the trade and industrial activities of the region. So, the processing units of North East India should prepare to avail this opportunities.

9. Technological improvements in packaging, processing, quality control, transportation, preservation technique and farm practices have led

to a rapid development of the fruit processing units in India. Hence, it is suggested that the existing fruit processing units of this region should go for modernisation of various methods of production such as, processing machines and equipments, packaging, and quality control systems.

10. Fruit processing plants need to go for strong backward linkages with the producing farms to ensure quality of production. Also to look out for a stable forward linkages with the market centres so as to survive in this era of cut-throat competition.
11. Technological upgradation at par with the best to achieve low cost of production, better quality and economies of scale will be required specially to meet the challenges arising out of WTO regime.
12. Fruit processing of North East India should study the demand of size, colour, quality and nutritional value of products according to the demand made by national market in general and foreign market in particular.
13. A North East Quality mark needs to be instituted to give proper incentive to scientific farming. Thus the low quality produces will be driven out of the market and the overall quality consciousness of the region will improve.
14. It is suggested that the state should go for the development of one particular fruits say pineapple so that the state may go for mass processing of pineapple.
15. The edible portions in most of the fruits vary between 50-70 per cent. Instead of transporting fresh fruits like pineapple, citrus, peach, pear, plum over long distance specially when connecting roads to villages and districts of the North Eastern Region are very poor, it may be more appropriate to carry pulps and concentrates. This will save transport cost and also

help in the economic exploitation of residue into value added by-products.

Hence, the overall scenario emerging after the analysis is that the fruit processing sector in the region presents an attractive opportunities for investment despite the poor performance of the existing 61 units in the region. However, this must be preceded by proper planning and a focused appeal towards the market requirements especially for an export oriented unit in the region.

Directions for Future Research

Any agenda for future research on fruit processing industry should contain the broad out line of problems relating to the external and internal factors affecting the fruit processing units, the kind of research required to be initiated, methodological refinements and the assistance to be extended to researchers in conceptualizing and developing appropriate research frames. The research on fruit processing industry would do well to have an empirical focus.

An effort has been made to analyse the emerging trends of fruit processing industry in North Eastern Region. Keeping in view the importance of the topic under study the following agenda for future research is suggested :

- Impact of WTO on fruit processing industry.
- Personnel Management in fruit processing industry.
- Marketing management in fruit processing industry.
- Financial management in fruit processing industry.

To sum-up, the research on fruit processing industry need greater deepening and widening . Its accent requires a shift from descriptive and narrative compilations to conceptual and policy transformations. The research needs a high degree of sophistication. There is a need to initiate researches on contemporary issues in the development of fruit processing sector.

Bibliography

Books

1. Aggarwal, Y.P : *Statistical Methods-Concepts, Applications and Computation* (Second Revised Edition), Sterling Publishers Private Ltd. New Delhi, 1990.
2. Batra, G.S : *Development of Entrepreneurship*, Deep & Deep Publications, New Delhi, 2002.
3. Cherunilam Francis : *Business and Government*, Himalaya Publishing House, Bombay, 2001.
4. Datta Ray B. & Prabin Baishya : *Sociological Constraints to Industrial Development in North East India*; Concept Publishing Company, New Delhi, 1998.
5. Devendra Thakur : *Agro based Industries*, Deep and Deep Publications, New Delhi, 1998.
6. Dhirendra Nath Borthakur : *Agriculture of the North Eastern Region with special reference to Hill Agriculture,* Dee Cee Prakashan, Guwahati, 1992.
7. Elhance D.N. et al : *Fundamentals of Statistics*. Kitab Mahal Allahabad, 1998.
8. Ghosh, B.N. : *Scientific Method and Social Research,* Sterling Publishers Private Ltd. New Delhi, 1999.
9. Ghosh, S.P. : *Horticulture in North - Eastern India*, Associated Publishing Company, New Delhi, 1985.
10. Ghosh, P.K.: *Business policy - Strategic Planning & Management*; Sultan Chand & Sons, New Delhi, 1994.

11. Girdhari Lal, G.S. et al : *Preservation of Fruits and Vegetables*, Publications and Information Division, ICAR New Delhi, 1998.

12. Goswami P.C. : A note on Development of Small Scale Industries in North East India in D.D. Mali & P. Deka, *Industrial Development in North East India*, Omsons Publication, Guwahati, 1987.

13. Gopal Swaroop : *Advances to Small Industries and Small Borrowers*, Sultan Chand & Sons, New Delhi, 1992.

14. Gupta. C.B. : *An Introduction to Statistical Methods*, Vikas Publishing House Pvt. Ltd., New Delhi, 1992.

15. Gupta, S.P. : *Statistical Methods*, Sultan Chand & Sons, New Delhi, 1991

16. Herman Steen : *Cooperative Marketing- The Golden Rule in Agriculture*, Double Day Page & Company, New York, 1923.

17. Hingorani, N.L. et al : *Management Accounting*, Sultan Chand & Sons, New Delhi, 1997.

18. Kothari, C.R. : *Quantitative Techniques*, Vikas Publishing House Pvt. Ltd. 1984.

19. Kothari, C.R. : *Research Methodology, Methods and Techniques*, Wishwa Prakashan, New Delhi, 1990.

20. Maheshwari S. N. : *Advanced Accountancy*, Vikas Publishing House Pvt. Ltd., New Delhi, 1994.

21. MS - II, : *Corporate Policies and Practice*, IGNOU study material, New Delhi.

22. Pareek U and T.V. Rao : *Developing Entrepreneurship*, *Learning Systems*, New Delhi, 1978.

23. Phillip Kotler : *Marketing Management — Analysis, Planning Implemetation and Control*, Prentice Hall of India, New Delhi, 1994.

24. Pandey, I.M.: *Financial Management*, Vikas Publishing House, New Delhi,1997.

25. Prasad L. M.: *Business Policy and Strategy,* Sultan Chand & Sons, New Delhi, 1995.

26. Ramaswamy, V.S. and Namakumari, S. : *Marketing Management; Planning Implementation and Control,* Macmillan India Ltd. Delhi, 1991.

27. Rao, K.V. : *Research Methodology in Commerce and Management.*

28. Rehman, A.R.M. : *Infrastructure Development for Industries,* Dibrugarh, 1997.

29. Shukla, S.P. and Aggarwal, A. K. : *Agriculture in North Eastern Region,* National Publishing House, New Delhi, 1986.

30. Srivastave, S.B. : *A Practical Guide to Industrial Entrepreneurs,* Sultan Chand & Sons, New Delhi, 1995.

31. Suman Bhatti & Uma Verma : *Fruit and Vegetable Processing Organisations and Institutions.*

32. Swaminathan, M.S. : *Biotechnology in Agriculture,* Macmillian India Limited, Madras, 1992.

33. Vasant Desai : *Dynamics of Entrepreneurial Development and Management,* Himalaya Publishing House Bombay, 1995.

Unpublished Works

Ph. D. Thesis

34. Chabungbam Jadunanda Singh : Structure and Growth of India's Financial Services. M.U. 2000.

35. KH. Kamala : Diversification as a Corporate Strategy : The Indian Scenario; M.U. 1993.

36. Th. Biren Singh : A Perspective of Agricultural and Horticultural Development in Manipur during the plan period ; M. U. 1993.

37. Rajmani Singha : Management in Handloom Indusry : A

Study of the Production and Marketing of Hanldooms for Export. M.U. 1992.

M. Phil. Dissertation

38. Nurun Nesha : The Development of Horticulture in Manipur during the plan period ; M.U. 1989.

Government Publications

39. Agricultural and Processed Food Products Export Development Authority (APEDA), Ministry of Commerce, Government of India.

i) APEDA QUARTERLY information

- Issue July 1999 APEX update
- Issue October 1999 APEX update

ii) Financial Assistance Schemes, APEDA

40. Centre for Monitoring Indian Economy,(CMIE) Mumbai.

i) Food and Beverages, Monthly Review of Investment Projects 2000.

41. Directorate of Commerce and Industries, Government of Manipur.

i) Industrial Policy of Manipur.

ii) Seminar on Development of Women in Food Processing Industries 13th – 14th September '99.

iii) Souvenir on Development of Food Processing Industries in Tribal areas of Manipur 1992

42. Directorate of Economics and Statistics, Government of Manipur.

i) Economic Review of Manipur, 1994-95.

ii) Statistical Handbook of Manipur, 1997-98.

43. Directorate of Horticulture and Soil Conservation, Government of Manipur.

i) A pamphlet on Mag-Fruit preservation factory.

ii) Draft Annual plan.

iii) Farmer's Training on "Integrated Development of Fruits"

44. Indian Council of Agriculture Research (ICAR) New Delhi.

i) INDO-USAID Sub-project (1985-1991) on Post Harvest Technology of Fruits and Vegetables.

ii) Technological options for North Eastern Hill Region, ICAR Research Complex for N.E.H. Region Umiam, Meghalaya.

iii) Vision 2020, ICAR Research complex for NEH Region, Perspective Plan. Umaim, Meghalaya.

45. Ministry of Commerce of Industry, Government of India, New Delhi.

i) Hand-book of Industrial policy and statistics 2001, Office of the Economic Advisor.

46. Ministry of Food Processing Industries, Government of India.

i) Food Processing Industries, Outline of Assistance Under Schemes of the Ninth plan 1997-2002.

ii) Ministry of Food Processing Industries, Annual report-1990-91, 1991-92, 1992-93, 1993-94, 1994-95, 1995-96, 1996-97, 1997-98, 1998-99, 1999-2000, 2000-01, 2001-02.

47. Ministry of Home Affairs, Government of India, North East Newsletter, A monthly Newsletter on the North Eastern Region of India.

i) Volume 1. No. 6. May, 2000

ii) Volume 2. No. 9. September 2001

iii) Volume 4. No. 7. July, 2002

iv) Volume 4. No. 8. August, 2002.

48. National Bank for Agriculture and Rural Development (NABARD), Mumbai.

i) Annual Report 2000-2001.

ii) National Food Processing Policy.

iii) Union Budget 2001-2002.

49. National Cooperative Development Corporation (NCDC), New Delhi.

i) Annual Report 1999-2000.

ii) Annual Report 2000-2001.

iii) J.N.L. Srivastava : Integrate the National Market for strong Agriculture Growth NCDC Bulletin, April, 2002.

50. National Horticulture Board (NHB) Ministry of Agriculture, Government of India.

i) Production Yearbook.

ii) Salient features of the National Board.

iii) Techno Economic feasibility study of the Development of Horticulture in the State of Manipur, 1994.

51. North Eastern Council, Shillong

i) A study on Packaging and Transportation Methods for the Horticultural Products of the North Eastern Region, Centre Food Technological Research Institute, Mysore – 570013, March, 1978.

ii) Feasibility report for revival of a Fruit Processing Unit at varengte (Mizoram).

iii) North Eastern at a Glance.

iv) Project report on Fruit Juice Concentration, Consultancy Cell, Food and Nutrition Board.

v) Techno-Economic Feasibility Report for a Fruit Processing Unit in Assam.

52. North East India Data Bank (NIC Assam State Unit, Guwahati).

i) Brochure.

ii) Industrial Policy for North Eastern Region.

iii) Industrial Policy of Assam 1997.

iv) Market Information on Essential Commodity (1998-99).

v) Market Information of Horticultural Product.

vi) Marketing Organization in the North Eastern States.

vii) Technologies for industries viable in the North Eastern Region.

53. Small Scale Sector : Hand-book of industrial policy and statistics 2001, Office of Commerce and Industry. GOI, New Delhi.

54. Statitical Organization, Government of India

i) Census of India, Series 14, 2001, Manipur.

ii) Selected Socio-Economic Statistics of India, 2000.

JOURNALS

55. Aggarwal, P.K. : Horticulture sector in India – Latest Status and Major Governmental Policy Interventions, *Political Economy Journal of India,* Centre for Indian Development Studies, Chandigarh Vol. II issues 1 & 2, January-June 2002.

56. Amarnath Bandarla: Problems of Agro-based Industries, *Khadi gramodyog.* The journal of rural economy. Volume XXXVII No. 5. February 1992.

57. B.K. Kumbar & B.P.N. Singh : Ways to Increase Profit Margin or Peduce Product Cost in a Food Industry. *Financing Agriculture,* volume XXIV No. 4. October - December 1992, p.p. 9-14.

58. C. Kurunanithy, M.S. Sajeev & R. Kailappan : Cold Storage-need and planning, *Kissan World.* Volume 25. No. 9, September 1998.

59. Dolly Jain, et al : Food processing industries in India, Export problems and prospects, *Indian Journal of Marketing* Y-21, Hauz Khas, New Delhi-110016. Volume XXXII November 8, August 2002.

60. Dubey Sanjeev : Food Processing Industries in India export

problems and prospects *Indian Journal of Marketing,* Volume XXXI, November 1-2 January-February 2001.

61. Er. P. Hemasankari & Er. S. Bhuvaneswari : Food Process Engineering, Need in India, *Kissan World,* August 1999, volume 26. No. 8

62. Gupta, S.P. : Possibilities of Agriculture Development through Cooperative Finance : An analytical study. *Financing Agriculture,* Agricultural Finance Cooperation Ltd. Bombay, volume XXIV No. 3, July-September 1994, p.p.8-13.

63. Kansera, S.L. : Looking closely at the various Segments of Modern Food Processing Industry in India. *Political Economy Journal of India,* volume II Issues 1-2, January-June 2002, Chandigarh.

64. Karmakar, K.G. : Capital Formation in Agriculture and Role of Institutional Credit, *Journal of Rural Development,* Hyderabad Vol. 17 No. 1, January-March, 1998.

65. Khanna Perminder : Horticulture Processing Problems and Prospects, *Political Economy Journal of India.* vol. 9 No. 115, 2000.

66. Mathur, B.C.: Cooperative in Marketing, Processing and Export of Horticulture Produce, *Productivity,* New Age International (P) Ltd. New Delhi, vol 36. No. 4. January-March 1996.

67. MH. Bala Subrahmaya: Reservation Policy for Small Scale Industry. Has it delivered the goods ? *Economic and Political Weekly* May 27, 1995.

68. Meera Mohadeen, P. M. : WTO Impact on Food Processing Industry, *Industrial Herald,* March 2002.

69. N. Srinivasam, et al: Consumer's perception towards processed fruit and vegetable products. *Indian Journal of Marketing* (New Delhi) volume XXX No. 11-12, November-December, 2000.

70. Prakash Anand: Post Harvest Experience in Agricultural

Marketing in Haryana. *Political Economy Journal of India,* vol. 9, Issues 1&2, January-June 2000. ISSN 0971 - 2097.

71. P.G. Adsule : Marketing and Export of Mango and its products. *Laghu Udyog Samachar,* January-December 1997.

72. P.G. Adusle : Post Harvest Technology of Grapes in India. *Laghu Udyog Samachar,* April-June 1999, vol. XXII, No. 9-11.

73. Raghunandan, A.: Diversifying Agriculture by Boosting Horticulture. *Financing Agriculture,* Agriculture Finance Corporation Ltd. Bombay, volume XXIV, No. 4, October-December 1992, p.p. 4-8.

74. Sarkar, A.N.: Post Harvest Management of Horticultural Produce. *Financing Agriculture,* Agricultural Finance Cooperation Ltd. Bombay, vol. XXIV, No. 1, January-March 1994, p.p. 8-17.

75. S.B. Maini & T.C. Anand : Fruit and Vegetable Processing Industry. Present status & future prospects, *Productivity* vol. 36, No. 4, January-March 1999, New-Age International (p) Ltd. publishers, New Delhi.

76. Singh H.P. : Existing status of marketing information system for Agricultural produce in India and need for modernisation. *Kissan World,* Coimbatore, March 1998, vol. 25, No. 3.

77. Singh, H.P. : Post Harvest Technology of Mangoes, *Financing Agriculture,* Agricultural Finance Cooperation Ltd. Bombay, vol XXVI, No. 3, July-September 1994, p.p. 18-23.

78. Sinha Anup: Economic Restructuring Transfer and Human Resource Development. *Dicision,* vol. 26, No. 1-4, January-December 1999.

79. S.S. Khanka : Small enterprises in international business, *Journal of Commerce,* Golden Jubilee volume, Guwahati University (1999).

80. Subrahmanyam, K.V. and Ghosh, S.P. : Horticulture

Development in Karnataka-past, present and future, *Indian Horticulture,* April-June 1999.

81. State Bank of India, Monthly Review Volume XXXIX No. 2, February 2000.

82. Sudha, S. : What would be the appropriate form of industries for the Indian situation ? The Journal of Rural economy volume XXXXV, No. 9. June 1999, Mumbai.

83. Suresh K. Chadha : Prospects of Agro food processed Industry. Political Economy Journal of India. July-December 1999, Vol. 8, No. 2. Issues 3&4. Journal of Centre for Indian Development Studies.

84. Susanta, K. Roy : Development of infrastructural facilities to boost export of fresh and processed Horticultural produce, *Financing Agriculture,* Agricultural Finance Corporation Ltd. Bombay, Vol XXIV, No. 4, October-December 1992. p.p.12-27.

85. Vipla Chopra : Food Processing Industry in India, *Political Economy Journal of India,* Centre for Indian Development Studies, Chandigarh, Vol. II Issues 1 & 2, January-June 2002.

86. V.K. Sehgal & M. Arora : Post Harvest Technologies for Rural Entrepreneurship through Agro-Industrial Complexes. *Political Economy Journal of India,* Vol. 9, No. 127.

87. Yojana, A Development Monthly

Volume 38 No. 24, January 15, 1995.

Volume 38 No. 24, November 1998

Volume 38 No. 24, December 1998.

Volume 44 No. 10, October 2000

Volume 45 No. 1, January 1

Volume 45 April 2001

Volume 45 September 2001

Volume 45 November 2001

Volume 46 February 2002.

NEWSPAPERS

The Sangai Express.

88. Micro finance scheme - NEDFi January 2000.

89. Nagaland to constitute three task forces, February 22, 2000.

90. Food processing policy. February 25, 2000

91. Budget highlights : Technology Mission for N-E Horticulture, February 29, 2000.

92. Yashwant Sinha : A blessing in disquise for poor states Union Budget 2000-2001. March 1, 2000.

93. H. Lokhon Singh : State budget 2000-2001, A budget for employees. March 14, 2000.

94. H. Lokhon Singh : Budget speech 2000-2001. March 14, 2000.

95. A.C. Sharma : Crisis in Manipur Agriculture, April 26, 2000.

96. Th. Chaoba : Food processing training centre to be set up. May 7, 2000.

97. Engam Pame : Impact of Development in the hills. May, 8. 2000.

98. Cold storage to be commissioned soon, May 15, 2000.

99. More aid for food processing units in difficult areas, May 26, 2000.

100. Leitongbam Indreshore : Magfruit production hits rock bottom, May 31,2000.

101. Need to develop agro-based industries to create employment avenues. June 4, 2000.

102. Government plants development of eco-village. June 5, 2000.

103. Chaoba seeks separate Ministry. June 6, 2000.

104. Nipamacha distributes power tillers to hundred farmer in all districts to own cold storage. June 15, 2000.

105. Chaoba moots food processing research centre in Imphal. June 15, 2000.

106. Chaoba to flow fund for food processing. June, 16, 2000.

107. FINER : Centre's N-E Industrial policy a failure. June 19, 2000.

108. Biotech fails to help poor farmers. July 12, 2000.

109. More flavored ginger oil for spice lovers. July 12, 2000.

110. B.P. Hazarika : Prospects and economics of mushroom cultivation in Manipur. July 13, 2000.

111. Salam Rajesh : Experiencing a market change. August 6, 2000.

112. Chaoba : Setting up of food processing units in Eastern Region. August 20, 2000.

113. Sam Dutt : Hi Tech Horticulture in India. September 20, 2000.

114. B.P. Hazarika : Possibilities of Industrial crop in Manipur. September 24, 2000.

115. Greater role for biotechnology in India urged. November 29, 2000.

116. Laisharam Shamungou : Thangal orange production faces grim future. November 30, 2000.

117. Chaoba : Plan for better food preservation. December 2, 2000.

118. Amar Yumnam : Manipur 2000. Time to sow, time to reap (A note for the perspective planning group, State Planning Board, Manipur) February 4, 2001.

119. R.K. Ranjan : Development of transport & communication system in Hills districts of Manipur. March 4, 2001.

200. Passion fruit cultivation at Mao. March 16, 2001.

201. Swaminathan foundation for sustainable food security. July 18. 2001.

202. ICAR : Agri-Technology should focus on local situations. October 1, 2001.

The Assam Tribune

203. Bhuyan Aradhana : The Lemon has several medicinal properties, October 31, 999.

204. Dreaded drinks, October 9, 1999.

The Nagaland Page

204. Brain drain - Consuming Vegetables contaminated with heavy metals and pesticides can lower a child's intelligence - January 6, 2000.

205. Steps to develop trade potential of North East. January 6, 2000.

The Telegraph

206. Fruit processing unit to be revived, proposal on pineapple project sought from Industries Directorate. September 8, 1998.

207. Jane E. Brody : Fruit juice, or the whole fruit ? March 15, 1999.

208. Contract farming set to begin in North Bengal. October 19, 1999.

209. Indronil Roy Chowdhury : India's industries policy must take into account hard truths about closure. October 20, 1999.

210. 7 pt. Assocham plan for 5 % farm growth. October 20, 1999.

211. Fruits to be exported from North East. November 5, 1999.

212. Farmer's market. November 15, 1999.

213. Gajinder Singh : Farmers turn shopkeepers in profit hunt. April 24, 2000.

214. Pushpa Grimaji : How not to pay for stale food . April 27, 2000.

215. Chaoba : National policy for food processing industry on the anvil. October 3, 2000.

216. N.E. Industrial Policy Review Underway ; December 3, 2002.

EMPLOYMENT NEWS

217. Dr. S. K. Sharma : Sustainable Development : Some key issues. August 1-7, 1998.

218. Highlights of the union budget 1999-2000. March 6-12, 1999.

219. Dr. Rajala Chandran : Women Empowerment in Agriculture. March 20-26, 1999.

220. Kumari Priyanka : Natural resource management towards sustainable agriculture in 21st century. February 12-18, 2000.

221. Nagaz Rasul : Food processing, value addition and income distribution. vol XXVI, No. 32. November 10-16, 2001.

222. Dr. A.P.J. Abdul Kalam : India Millennium Missions 2020 challenges. November 3-9, 2002.

BUSINESS STANDARD

223. Food processing industry in West Bengal. March 12, 2001.

224. Ruba Zachariah, Rasna maker piona, del monte tala Jv. March 12, 2001.

THE HINDU SURVEY OF INDIA

225. N.N.S. : Processed foods value addition holds the key, 1994.

226. Food Processing - A fast growing business, 1994.

OTHERS

227. Dr. Rajmani Singha & A.S. Yarso ; Agro-Based Industries

for Food Security in Manipur : with special reference to problems and prospects of fruits and vegetables processing industries. Abstracts of papers presented on National Seminar on Food Nutrition, Food Security, Public Distribution System in North-East India. 27-28 November, 2000, NECSSR. Shillong.

228. Dr. S.V. Ngachan : Agricultural Technology for food security in Ukhrul District, Seminar at Ukhrul, November 1998.

229. M. Tonjao Singh : Fruit and Vegetable Processing Industries in Manipur, Seminar at Imphal, June, 2000.

300. Proceedings of the National Seminar on Food, Nutrition, Food Security, Public Distribution System in North – East India, NEICSSR, Shillong, 27-28 November, 2000.

301. Proceedings of the two days Business Summit of the North Eastern States in Mumbai (19-20, July 2002) organized by Ministry of Development for North Eastern Region.

302. Singh R.K.P. et al : Post Harvest Management of Fruits and Vegetable in North East Region, Seminar at Shillong, 1998.

Abbreviations

APEDA	-	Agricultural and Processed Food Products Export Development Authority.
ARDC	-	Agricultural Refinance and Development Corporation.
CFTRI	-	Centre for Food Technological Research Insititute.
CWC	-	Central Warehousing Corporation.
DICs	-	District Industries Centres.
EOUs	-	Export Oriented Units.
FMCG	-	Fast Moving Consumer Goods.
FDI	-	Foriegn Direct Investment.
FAO	-	Food and Agriculture Organisation.
FPO	-	Fruit Products Order.
F&VP	-	Fruit and Vegetable Processing
FCI	-	Food Corporation of India.
HT	-	Hurdle Technology
IEMS	-	Industrial Entrepreneurial Memoranda.
ITDP	-	Integrated Tribal Development Programme.
IID	-	Integrated Infrastructure Development.

IIHT	-	Indian Institute of Handloom Technology.
ICAR	-	Indian Council of Agriculture Research.
IT	-	Information Technology.
KW	-	Kilo Watt.
MFPI	-	Ministry of Food Processing Industries.
MATA	-	Management of Tribal Area.
MODVAT	-	Modified Value Added Tax.
MN-I	-	Mini Mission - I.
MN-II	-	Mini Mission - II.
MN-III	-	Mini Mission - III.
MN-IV	-	Mini Mission - IV.
MW	-	Megawatt.
MT	-	Metric Tonne
NGOs	-	Non-Government Organisations.
NEDFi	-	North East Development Financial Corporation.
NHDC	-	National Handloom Development Corporation.
NCDC	-	National Bank for Agriculture and Rural Development.
NABARD	-	National Co-operation Development Corporation.
NERAMAC	-	North Eastern Regional Agricultural Marketing Cooperation Limited.
NAFED	-	National Agricultural Cooperative Marketing Federation of India.

NEC	-	North Eastern Council.
NEITCO	-	North Eastern Industrial and Technica Consultancy Organisation Ltd.
PMRY	-	Prime Minister Rozgar Yojana.
PSUs	-	Public Sector Units.
R&D	-	Research and Development.
RBI	-	Reserve Bank of India.
SIDBI	-	Small Industrial Development Bank of India.
SLC	-	State Level Committee.
SSI	-	Small Scale Industry.
SWOT	-	Strength Weakness Opportunity Threat.
TCS	-	Tata Consultancy Services.
VSI	-	Village Small Industries.
WTO	-	World Trade Organisation.
WSCs	-	Weavers Service Centres.

Appendix - I

Fruit and Vegetable Processing Units in the North Eastern States of India as on November, 1999.

SL. No.	Name	Capacity	Products Name	Brand	Present status
Assam					
1.	M/S. United Soft Drinks Ltd. Habogaon, Jorhat		Sweentend Aerated Water		
2.	M/S. North East Pure Drinks (P) Ltd. Rani, Guwahati		Sweented Aerated water Fruit R.T.S		
3.	M/S. Assam Hills Small Industrial Development Corporation Ltd. Diphu,Karbi-Along	1.5 t/d	Juice, Jam Jelly, Squash	KANCH	Not Functioning
4.	M/S. Fulertal Agricultural Products Processing & Marketing Co-operative Ltd., Cachar	1.5 t/d	Pineapple juice, jelly		Not Functioning
5.	**M/S. Modern Food Processing Co-operative Society Ltd. Unit-II, Bonda Industrial Complex, Guwahati.**	0.5 t/d	Pickles, Squash Chutney, Jelly Jam.	NIRU'S	Operatinal
6.	M/S. North Eastern Fruit & Vegetable processing Industry Baladmari, Goalpora.		Juice, Pickles Squash, Jam, Sauces		

7.	M/S. B.K. Gees Food preserves, Niva market Paltan Bazar, Guwahati.	0.3 t/d	Jam, Squash, Pickles, Sauces	GEES	Converted into bakery Business
8.	M/S. Hindustan Food & Chemical Industries, Silchar.		Squash, Sharbats		
9.	M/S. Nectar Benverages (P) Ltd. Beltola, Guwahati		Sweented Aerated Water		
10.	M/S. North Eastern Fruit Tusar Complex, Guwahati.		Squashes, Pickles, Fruit RTS, Souces		Not Functioning
11.	M/S. Paul's Fruit Products, Silchar.		Squash, Fruit RTS		
12.	M/S. New Paul's Fruit Products, Silchar		Squash, Jam Synth, Syrups		
13.	M/S. Arati Fruit Product Aryyapatty, Silchar.		Squash Synth Vinegar		
14.	M/S. Mitra Enterprises Mahaprabhu Sarani, Silchar		Jam, Squash		
15.	M/S. Indu's Fruit Product Juria, Nagoan.	50 t/yr	Squash, Pickles	TRIPTA	
16.	M/S. Biswandas Food Processing Works, Guwahati.	50 t/yr	Pickles Squash		
17.	M/S. Satyam Food Products, Sonitpur.		Pickles Chutneys		
18.	M/S. Digo Food Products, Dispur, Kamrup.	0.5 t/d	Squash, Pickles Sauces		
19.	M/S. J.M. Food Products, Nagaon.	12t/yr	Squashes Pickles	SUSWAD	Operatinal

20.	**M/S. Deepashree Industries North Lakhimpur.**		Juice, Squash Jam.	DEEPA'S	Operatinal
21.	M/S. Shri Ganesh Products Zoo-Narangi, Guwahati.		Pickles		
22.	M/S. Shanti Sadhan Asharm, Guwahati		Jam, Squash Pickles, Sauce		
23.	**M/S. Meghali Fruit Processing, Lakhimpur.**		Juice, Squash Jam.	MAJUDEE	Operatinal
24.	M/S. Assam Horticultural Research Centre, Guwahati.		Pickles		
25.	M/S. North Eastern Agricultural Marketing Corporation Ltd., Guwahati.		Squash, Juice Pickles, Jam.	NERAMAC	Relabeller
Manipur					
1.	**M/S. The Manipur Fruit Processing & Cold Storage Co-operative Society Ltd. Imphal.**	3.M.T/d	Squash	HEIRANG	Operatinal
2.	**M/S. Govt. Fruit Preservation Factory, Imphal.**	0.75 M.T/d	Squash Jam	MAGFRUIT	
3.	**M/S. Waifruit, Thoubal**		Squash Pickles	WAIFRUIT	Operatinal
4.	M/S. The Universal Wonder-Fruits, Lamsang Bazar, Imphal.		Squash, Juice pickles	IMA	Operatinal
5.	**M/S. Manipur Fruit Juice, Crusher, M.G. Avenue, Imphal.**		Juice Squash	MANIFRU	Operatinal
6.	M/S. Ratna Fruit India, Uripok Polem Leikai, Imphal.		Juice, Squash RTS, Ice Candy	RATNA	Operatinal

7.	M/S. Juicy Food Industries Porompat, Imphal.			LIPSIP	Operatinal
8.	M/S. Evergreen Fruit Preservaton Works, Canchipur, Imphal.	200 kg/d	Juice Squash	AMRITA	Not Functioning
9.	M/S. Manipur Hill Fruits Processor, Pungdongbam, Imphal.	0.2 t/d	Jam, Squash Fruit RTS Pickles	CHANDNI	Operatinal
10.	M/S. A Family of Food Products, Imphal.		Squashes, Soft Drinks, Jelly	SANA	Operatinal
11.	M/S. Thangjam Agro-Industries, Chingmeirong, Imphal.	17.5 MT/yr	Squashes	LIKLA	Operational
12.	M/S. Amar Foods & Packaging co. (P) Ltd. Wangkhei, Imphal.		Fruit R.T.S.		
13.	All Manipur Food Preservers Association, Imphal.			MAAPY	Relabeller
	Arunachal Pradesh				
1.	M/S. Arunachal Horticultural Processing Industries Ltd. Nigmoi West Siang.		Juice, Squash Pickles, Sauce		
2.	M/S. Fruits & Vegetables Processing Centre, Regional Apple Nursery, West Kameng.		Squash, Jam Pickles		
3.	M/S. District Horticulture Officer, Government of Arunachal Pradesh, Along, West Siang.		Jam, Squash		

Meghalaya			
1. M/S. Associated Beverages (P) Ltd. Industrial Area, Burnihat.	Sweetened Aerated water.	GOLDPINE	Operatinal
2. M/S. ATC Agro Industries Ltd. Burnihat, Re-Bhoi			
3. **Govt. Fruit Preservation Factory, Fruit Garden Shillong.**	Juice Squash	MEG	Operatinal
4. M/S. Fruit Preservation Officer, Dainadubi East Garo Hills.	Squash, Jam		
5. **M/S. Hibi Fruit Products, Krishnamandir, Jail Road, Shillong.**	Squash	HIBI'S	Operatinal
6. M/S. Meghalaya Orange Honey (P) Ltd. Nigkerm, Shillong.	Squash		
7. M/S. Raja Fruit Products, Lower Jail Road, Shillong.	Squash		
8. M/S. Loca Food Products Nongthymai, Shillong.	Squash Pickles		
Nagaland			
1. M/S. Nagaland Beverages (P) Ltd. New Industrial Complex, Rangapahar, Dimapur.	Sweetened Arated Water		Operatinal
2. M/S. Govt. Fruit and Vegetable Processing Plant, Viola Colonmapur, Dimapur.	Juices, Squashes Pickles R.T.S.		
3. M/S. Fruit Canning Factory, Longnok Changki, Makokchung.	Jam, Squash Pickles Squash		

4.	M/S. Chakhesang Women Multipurpose Co-operative Society, Pfutsero, Kohima.		Jam, Squash Pickles		
5.	M/S. Limha's Fruit Products, New Field Colony, Dimapur.		Squash		
6.	M/S. Immanuel Products, Rongesnmongal, Mokokchung.		Jam, Squash Pickles		
Tripura					
1.	M/S. Tripura Small Industrial Corporation Ltd. Agartala, Tripura.		Canned fruits Jam, Squash		
2.	M/S. North Eastern Regional Agricultural Marketing Corporation Ltd. Fruit Juice Concentration Plant, Nalkata, Tripura.		Fruit Juice concentrate		Operatinal
3.	M/S. M.D.S. Home Products, Agartala, West Tripura.		Jam, Pickles		
4.	M/S. Centre for Social Work, Melar Math, Agartala.		Squash, Jam Pickles		
Mizoram					
1.	Fruit Juice Concentrate Plant, Chhing-Chhip MIFCo Ltd. Aizawl, Mizoram.	65.8 % C.U.	Juices, Squash Jam	MIFCo	Operatinal
2.	M/S. Zoram Food Industries, Aizawl, Mizoram.		Squash cordials Pickles.		

Note : The bold letter units were selected for the present study.

Appendix - II

Provisional List of Food Processing Industrial Units in Manipur

Type of Units	*Imphal*	*Thoubal*	*Bishn-upur*	*Churach-andpur*	*Tameng-long*	*Sena-pati*	*Chandel*	*Ukhrul*	*Total*
Rice Mill	1153	175	119	251	61	60	26	271	2116
Oil Mill	4	-	2	-	-	2	-	-	8
Flour Mill	2	-	-	-	-	-	-	-	2
Dal Mill	44	3	3	2	-	7	-	1	60
Spices	7	-	-	3	-	-	1	-	11
Potato Chips	9	-	-	-	-	-	-	2	11
Bakery	46	-	4	23	2	5	1	44	125
Iodised salt	2	-	-	-	-	-	-	-	2
Fruit processing	12	1	-	-	-	-	-	-	13
Chow-chow	4	-	-	-	-	-	-	-	4
Bee keeping	-	-	-	-	-	5	-	-	5
Ginger processing	1	-	-	-	-	-	-	-	1

(Contd...)

Cold drink	1	-	-	-	-	-	-	1	
Poultry feed	3	-	-	-	-	-	-	-	3
Poultry meat	10	-	-	-	1	-	-	-	11
Pop corn	8	-	-	-	-	-	-	-	8
Fish - fermentation	79	-	17	-	-	-	-	-	96
Soft drink	2	-	-	-	-	-	-	-	2
Confectioneries	1	-	-	-	-	-	-	-	1
Bambo shoot	3	-	-	-	-	-	6	-	9
Knol-knol	35	-	-	-	-	-	-	-	35
Total	1426	179	145	279	64	79	34	318	2524

Source : Directorate of Commerce and Industries, Government of Manipur.

Appendix - III

Statement Showing of Food Processing Industries Recommended to the Ministry of Food Processing Industries.

	Name of the project & Address.	*Particulars of the Proprietor*	*Name of the District in which the project is proposed*	*Total estimated cost of project Amount*	*Year of recomen-dation*	*Proposed grant-in-aid to MFPI, Govt. of India.*	*Promoter's contribution*	*Year of Approval*	*Sanction Amount*	*Remrks*
1.	Wate Fedes Speeces Ind., Khongjom Bazar, Manipur.	M. Hemanta Singh, Wangjing Tentha Farmer's Development Assn.	Thoubal District.	32,81,850	16-3-95	6,56,370	19,69,110	26-9-1995	12,78,000	The installation of machineries has already completed production is not yet started.
2.	Kyuma Potato Chips, Sega Raod, Imphal.	Y. Surendra Singh, Organiser Konjeng Hajari Leikai Youth Unemployed Members Assn. Sega Road, Imphal.	Imphal District	14,67,468	3-2-96	4,40,240	7,35,468	30-7-1996	5,50,000	The installation of machineries has not yet done.

(Contd...)

3.	Integrated Mini Cold Storage, Wangbal.	Kumar Singh, Secy. Integrated Rural Development & Educational Organisation, Wangbal, Thoubal District.	Thoubal District	51,84,218	3-2-96	25,92,195		21-3-1997	25,92,000	Construction of Building Completed and machinaries yet to be installed.
4.	Bishnu Fruits & Vegetable Plant, Bishnupur Bazar, Manipur.	Y. Biren Singh, Secy. Malom District Tuliyaima Dev. Assn.	Bishnupur District	32,54,432	18-11-96	16,15,460	16,17,216	11-3-1999	11,20,000	Construction of Building completed and machinaries yet to be installed.
5.	AFORA Chilli Processing Industry, Yangangpokpi, Ukhrul.	A.S. Athaan, Secy. Action for Rural District Dev., Yaingangpokpi, Ukhrul.	Ukhrul District	23,95,706	10-2-97	8,29,682				Sanctioned is awaited.
6.	Passion Fruit Processing Industry, Punamei, Mao.	L. Dikhe, President, TheSamaritan Social Service Assn. Mao.	Senapati District	15,11,831	18-11-96	7,55,916				Report yet to be submitted by the organisation.

(Contd...)

7.	Indo Burma Fruit & Vegetable Dehydration Plant, Thonju, Canchipur.	Dhanachandra Sanasam, Secy. Integrted Economic Dev. Social, Thongju, Canchipur, Imphal.	Imphal District	30,63,980	10-2-97	10,61,126				1st installement released and4.36 lakhs vide letter 4-20/97 F & VP Desk dt. 17-12-99.
8.	Rada Bamboo Processing Plant, Lukhambi, Noney, Tamenglong.	N. Tamei. Secy. Rural Area Dev. Assn. Lukhambi None, Tamenglong.	Tamengllong District.	37,11,500	10-2-97	13,20,000	9,69,200	4-2-99	10,17,000	The Association reply a letter No. 9-9-1998 to the querie of CFTRI, Mysore for setting up of Bamboo Shoot Processing Plant at None, vide Ministry's letter No. 4-21/1997-F&VP dt. 1st Sept., 1998 (sanctioned recently)
9.	New Fish Processing Industry, Langol, Imphal.	H. Arunkumar Singh, Secy. All Manipur Rural & Urban Dev. Services, Singjamei Sorokhaibam.	Imphal District	47,65,000	17-2-97	34,22,000		23-6-99	10,10,000	New Fish processing Plant is agreed to sanctioned by the Ministry of Rs. 17.99 lakhs recently.

(Contd...)

10. Khongjom Pineapple Processing Industry, Khongjom, Tekcham.	S. Suresh Singh, Secy. Educated Un-employed Youth Dev. Assn., Khongjom Tekcham.	Thoubal District	39,90,260	25-4-97	20,72,860	7,60,013	3-3-99		A Project proposal was forwarded to the Ministry on 25-4-97 for setting up of a Pineapple Prodessing Plant at Khongjom. Under consideration in the Ministry.
11. R. Tuime Lolley, Secy. Peoples Action for Rural Dev. Association, Kachai, Ukhrul.	R. Tuime Lolly, Secy. Peoples Action for Rural Dev. Assn. Kachai, Ukhrul.	Ukhrul District	25,33,376	16-8-96	13,48,000			8,51,000	All the necessary requirement of the Ministry has already been submitted (sanction recently)
12. National Food (based on fruit & vegetable product).	Kh. Loken Singh, Secy. Providence Multipurpose Co-operative Society Ltd. Sega Road, Imphal	Imphal District	30,00,000	10-6-97	17,73,000		31-3-98	14,43,000	The construction of new building for the project was already completed and the reqd. machineries has not yet been installed.

(Contd...)

13. N A F, Assembly Road.	H. Kangjamba Singh, Chairman.	Imphal District	8,97,700	29-3-97	4,34,400		4,34,4000	The Assn. has already prchase one 407 Mini Tata for transporation collection and ceiling office form various parts of Manipur ans also from outside the state Manipur. Utilisation certificate submitted already.
14. Tamei Area Fruit & Vegetable Growers cum Marketing Co. Society Ltd.	Z. Alai, Chairman.	Tamenglong District	6,00,000	9-10-98	6,00,000		6,00,000	Alredy got assistance for setting up of Food Processing Training Centre at Tamei without the recommendation of Nodal Agency.
15. Poultry Meat Canning Unit.	S. Rishikumar Singh, General Secy. Kangleipak Loina Sinlon Luptin.	Imphal District	137,22,000	30-5-97	98,43,400			

Source: Directorate of Commerce and Industries, Govt. of India.

Appendix - IV

Interview Schedule for Collecting Data on the Field Study of Fruit Processing Units in N.E. India: A Case Study of Manipur and Assam

Block - 1 General Information

1.1 Name and Complete address of the sample Fruit Processing Unit : ______________________

____

1.2. Registration Office : ______________________

1.3. Registration No.: ______________________

Date and year of registration : ______________________

1.4. FPO License No. : ____________ Date of issue ________

1.5. Date and year in which processing started : __________

1.6. Brand Name : ______________________

1.7. Objectives of the Processing Unit : ______________________

____

Block-II Production

2.1 What are the different types of Machine used for processing fruits ?

Sl. No.	Name of the Machine	Name of the Manufacturer	Year of Purchasing	Cost of Machine (Rs.)
1.				
2.				
3.				
4.				
5.				
6.				

2.2. What is the capacity of your machine (Per hour)

Sl. No.	Name of the Machine	Full Capacity (in tonnes/kg)	Present Capacity Utilisation (in tonnes/kg.)
1.			
2.			
3.			
4.			
5.			
6.			

2.3. Types of fruits used for processing : (1) ____________

(2) ____________ (3) ____________

(4) ____________ (5) ____________

(6) ____________ (7) ____________

2.4. Fruits are purchased from (Rs. Per kg.) :

Sl. No.	Name of Fruits	Farmer	Co-op Societies	Private Traders	Post harvest Contractors	Others
1.						
2.						
3.						
4.						
5.						
6.						

2.5. Name and place from where you bought the above fruits :

Sl. No.	Name of Fruits	Place from where you bought
1.		
2.		
3.		
4.		
5.		
6.		

2.6. Which factor influence you in buying raw fruits : (tick)

(a) Good Quality (b) Low price (c) Easy Availability

(d) All of the above (e) Any other; please specify________

____

2.7. Other ingredients used for processing are purchase from /supplied by (Rs.per kg.)

Items	Government	National Market	Total Quantity required in a year
1. Sugar			
2. Preservatives			
3. Colours			
4. Citrid Acid			
5. Pectine			
6. Other additives			

2.8. How many persons are employed in your processing unit?

Nature of work	Types	Sex				Children below		
		Male	Wages Per day Per worker	Female	Wages Per day Per Worker	14 yrs.	18 yrs.	Total
Full Time	Skilled							
	Unskilled							
Part Time	Skilled							
	Unkilled							

2.9. Total working days and working hours of your processing units :

Working days in a year		Working hours per day	
Minimum	Maximum	Minimum	Maximum

2.10. Average consumption of fruit per day :

Types of fruits	Quantity (kg./Ton.)
1. ______	______
2. ______	______
3. ______	______
4. ______	______
5. ______	______
6. ______	______

2.11. Specified the types of products produced per day :

Sl. No.	Fruits	Juices	Squash	Jame	Others (Unit)
1.		(Litres)	(Litres)	(kg.)	
2.		(Litres)	(Litres)	(kg.)	
3.		(Litres)	(Litres)	(kg.)	
4.		(Litres)	(Litres)	(kg.)	
5.		(Litres)	(Litres)	(kg.)	
6.		(Litres)	(Litres)	(kg.)	

2.12. Do you have separate quality control section in your factory to maintain quality of your product ? Yes/No, if yes why local product is unable to compete in National market ?

____

____

2.13. What are the best items of your production ?

Sl. No.	Items	Actual Cost (per unit)	Whole Sale price (per unit)	Maximum Retail price per unit
1.				
2.				
3.				
4.				
5.				
6.				

2.14. Is there any by-product manufactured by your unit ? Yes/ No, if yes, what is the name of that by-product ?

____

If 'No' how you treat the waste materials ______________

____

2.15. What are the general problems faced in the area of production :

a) Supply of raw materials

b) Quality of raw materials

c) Maintenance

d) Storage of raw mateials / finished product.

e) Any others, please specify.

2.16. How to overcome the above problems in 2.15. Please give some sugessions, if any ______________

Block - III Marketing

3.1. Do you conduct costomers survey before producing your product to know the customer's needs and requirements like price, quality and size ? Yes/No, If yes how often and how many of them are satisfied with your product ?

Market survey Per year :

Parameters	Price	Quality	Size
Satisfied	%	%	%

3.2. To whom your product is sold ? Local/outside.

If local, to whom ______________

If outside- a) to whom ______________

b) through whom ______________

3.3. Whether the local production of your unit is sold by cash or on credit. If credit what is the average time lag for collecting amount of credit from the customer ?

____

3.4. What are the channel of distribution do you follow ?

a) Producer → Consumer

b) Producer → Retailer → Consumer

c) Producer → wholesaler → Retailer → Consumer

3.5. Do you have your own sale depot ? Yes/No, If yes, where it is located.

____

3.6. Which methods of pricing do you follow to boost up the sale of your product ? (tick)

a) cost based pricing b) Demand based pricing

c) Competition based pricing d) Any others, please specify

____

3.7. Whether the total annual production is sold during the year : Yes/No. If no, how much quantity of your unsold product is lying in the godown every year. ____________

____

3.8. What are the tools used for promoting sales of your product. (tick)

a) Personal b) Advertising through different media.

c) Sales promotion d) Combination of all of the above

e) Any others please specify.-____________________

3.11 Do you think packaging is required for your product. Yes/ No. If yes, what is your packaging strategy ? (tick)

a) Plastic bottle b) Glass bottle

c) Canned d) Tetra packs

e) Tin f) Any others, please specify ________

3.10. Who is the supplier of your packaging materials. Local or outside.

(Please indicate the name of the firm) ________

____

3.11. Do you require transportation for sending your finished product to the distribution centre. Yes/No. If yes, how this requirement is carried out by (tick)

a) Owned vehicle b) Hired on contract basis

c) Hired on daily basis d) Any others, please specify.

____

3.12. Who is your keen competitor, local or outside firm. Please specify

a) the name of the firm and b) competing area like price, quality, size

a) ________ b) ________

3.13. Is there any participation of your product in exhibition at the State/National/ International level. Yes/No. If yes, mentioned the place and organised by whom ? (Latest)

1. ________

2. ________

3. ________

3.14. General problem in Marketing ________________

____

3.15. How to overcome the above problems mentioned in 3.14. Please give your suggestions. ________________

____

Block-IV Finance

4.1 How much amount of capital do you invest for your processing unit ?

Please specify categorically.

a) Fixed capital Rs.

b) Working capital Rs.

Total Rs.

4.2. How much amount of money is actually needed as working capital ?

Rs. ________________

4.3. Who funded the initial investment ? (tick)

a) by own sources

b) by financial institutions and commercial banks

c) Other sources.

Please specify the name of your funding agent.

4.4. Is that factory building your own or rented ?

a) If it is your own, how much amount of capital has been invested for the building ? Rs.________________

b) If it is rented, how much amount is paid per month as rent ?

Rs. ________________________________

4.5. Minimum times lag for converting raw material into cash__________________________________

____

days.

4.6. Do you get any Government subsidy, financial help etc. Yes/ No. If yes, mentioned the date, amount and the name of the scheme :

____

4.7. Are you satisfied with the amount of assistance provided by the government ? Yes/ No. How much do you expect? Rs. __________________________________

4.8. Do you satisfied your depreciation policy ? Yes/ No. if yes, then what method of depreciation do you adopt (tick)

a) Straight line method

b) Diminishing balance method /WDV method.

c) Annuity Method

d) Any other, please specify ____________________

4.9. Are the Government imposing tax on your business. Yes/ No, If yes, at what rate of ____________________

4.10. Give your opinion regarding the tax imposed by the Government (tick)

a) Too high b) Moderately high

c) High d) Low

e) Any others, please specify ____________________

4.11. General problems in getting finance and repayment ?

____

4.12. How to overcome the problem mentioned in 4.11. ? Please give your suggestions ____________________

(Please furnish Balance sheets of your unit for the last five years).

Block - V General Problems

5.1. Are you aware of the present industrial policy 1996 of the government of India with regard to Food Processing Industry. Yes/ No, if yes, then how will you justify in this regard ?

____

5.2. Are you aware of the present Industrial policy 1996 of the state government with regard to Food Processing Industry. Yes/ No., if yes then how will you justify in this regard ? (tick)

a) useful b) favour c) not applicable

d) any other please specify ________________

5.3. Are you getting adequate supply of power. Yes/ No. Whether it is available at subsidized rate. Yes/No.

Actual power available ( ) Power required ( )

5.4. Is there adequate transport facilities available in your areas?

Yes/No. Whether the transport subsidy is available to you from the state government and central government? Yes/ No. ____________________________________

5.5. Are you getting loan facilities from the banks and financial institutions ? Yes/ No. If yes, at what rate of interest

____

5.6. Do the government or other institutions impart trainings to the workers of your unit. If yes, menton the name of the training and year.

____

____

5.7. Give your generous views

a) About the government's attitude towards your firm ?

____

b) About the industries department's attitude towards your firm ?

____

5.8. Are you aware of the product competitiveness in the domestic/ regional/neighbouring market? Yes/No. If yes what would be the measure/your strategy to tackle with.

a) Combine with the fruit growers

b) Combine with the other Fruit Processing Units.

c) Combine with other food manufacturing units

d) any others, please specify ______________________

____

5.10. Any other problems which is not mentioned in the above.

____

____

5.11. How to overcome the problems mentioned in 5.10? Please give your suggestions ______________________

____

Signature of Investigator

Date ....

Signature of Respondent

Date ....

Appendix - V

Chi Square (χ^2) is calculated by dividing the squared difference between observed frequency (O) and expected frequency (E) and summing all these quotients for all categories. In symbols

$$\chi^2 = \frac{(O-E)^2}{\sum[E]} \quad \text{with degree of freedom (N-K)}$$

Where N = Total number of classes

K = Independent restrictions.

In simple words, Chi Square is a measure of the differences between observed frequencies and expected frequencis and throws light whether the difference between these frequencies is significant or not.

The test statistics for Kruskal-Wallis test which uses the above χ^2 test is given below:

$$H = \frac{12}{n(n+1)} \sum_{i=1}^{K} \frac{r_i^{\,2}}{n_I} - 3(n+1)$$

Where n = Total number of elements in K samples

H follows a χ^2 distribution with (K-1) degrees of freedom.

Prodedure for Testing a Hypothesis

1. Formualte the null hypothesis as follows :

   H_O = There is no difference in the levels of the processing units in generating the particular data.

2. Select the level of significance i.e., 5% or 1%.

3. Compute the test statistic using the above K-W test statistics on the basis of formulated null hypothesis.

4. Compare the calculated value with the table value of χ^2 at a specified level of significane for given degree of freedom.

5. Arrive at a conclusion as follows :

   i) Reject H_O (null hypothesis), if calculated value of H is greater than χ^2 (table value).

   ii) Accept H_O, if calculated value is less than table value.

Appendix - VI

Correlation Coefficient : Correlation is joint variation of two variables neither of which is restricted. Analysis of correlation enables us to determine whether an association exists between two variables and if it does, to what extent. Coefficient of correlation measures the degree of association between two variables. The two variables selected in this analysis are cost of production per unit and transportation cost per unit which can establish the degree of association for future actions.

In this primary years data for the above two factors of eight fruit processing units from Manipur are considered and the correlation coefficient is computed by using the Karl Pearson's correlation formulae. The purpose of using this technique is to establish whether positive correlation or negative correlation or spurious correlation exists between the above variables. Karl Pearson's Correlation formulae/equation used is :

$$r = \frac{\sum xy - n\bar{X}\bar{Y}}{\sqrt{\sum x^2 - n\overline{x^2}}\sqrt{\sum y^2 - n\overline{y^2}}}$$

Where Σxy = sum of the production of two variables x and y

x = mean of x variables

y = mean of y variables

n = No. of sample population.

Appendix - VII

Deference Test is applied in the case of paired data and in this context 't' is calculated as under :

$$t = \frac{\overline{X}\,\text{diff} - O}{\sqrt{O\,\text{diff}}} = \frac{\overline{X}\,\text{diff} - O}{\sigma\,\text{diff}}\sqrt{n}$$

Where $\overline{X}$ Diff or D = Mean of the difference of sample items.

O = The value zero on the hypothesis that there is no difference.

σ Diff = Standard deviation of difference and is worked out as

$$= \sqrt{\frac{\sum\left(D - \overline{X}\,\text{diff}\right)^2}{n-1}} \quad or \quad \sqrt{\frac{\sum D^2 - (D)^2 \cdot n}{(n-1)}}$$

Where D = Difference

n = Number of Pairs in two samples and is based on (n—1) degree of freedom.

Appendix - VIII

Market Survey to Know Total Volume of Sales per season

Shop No.	*Total Qty. (Bottles)*	*Local*	*Outside*	*Remarks*
1	*2*	*3*	*4*	*5*
Chingmeirong Market				
1	24	10	14	Kissan
2	81	72	9	Kissan
3	124	60	64	Kissan
4	48	12	36	Kissan
Deulaland				
5	100	70	30	Kissan
6	92	40	52	
7	12	8	4	
North BOC Point				
8	892	60	832	Top Choice Mandalay Myanmar Rs. 35-45 per litre profit margine 10-20% local profit margin Rs. 4 per bottle.
9	36	20	16	Kissan, Top Choice
10	100	28	72	Rap made in Thailand
11	40	30	10	
12	50	40	10	

1	2	3	4	5
13	80	20	60	
14	15	6	9	
Lamphel Sanakeithel				
15	24	18	6	
16	36	24	12	Kissan Individual preferred
17	13	5	8	
Super Market				
18	210	90	120	Kissan, Rallies (Calcutta)
RIMS Road				
19	100	17	83	
20	300	130	170	
21	135	60	75	Real-Dabur Rs. 68 per litre
22	100	30	70	
23	240	150	90	Kissan, Real
24	120	96	24	
Nagamapal				
25	300	100	200	
26	300	180	120	
Khoyathong				
27	60	25	35	Kissan

1	2	3	4	5
28	24	10	14	Kissan
29	36	—	36	Kissan
30	150	45	105	
31	20	5	15	
32	24	20	4	
33	15	5	10	Kissan
34	100	80	20	
35	48	18	30	Kissan, Rallies (Calcutta)
MG. Avenue				
36	120	106	14	
37	24	10	14	
38	50	20	30	
39	48	30	18	Pepsi Caputuring the market
40	36	20	16	Local cheaper
41	16	6	10	
Thangal Bazar				
42	78	30	48	Bikaji-Rajasthan Rs. 60 per litre
43	60	10	50	
44	60	50	10	
45	86	20	66	
46	200	80	120	
47	12	4	8	
48	100	40	60	

1	*2*	*3*	*4*	*5*
Uripok				
49	50	10	40	Outside brand like Kissan preferred
50	12	2	10	
51	50	30	20	
Singjamei				
52	72	40	32	Merry, Mandalay Myanmar Cheaper and hance preferred
53	180	—	180	Local Cheaper than outside product.
54	60	50	10	
55	60	40	20	
56	100	30	70	
Total	5523	2214	3309	

Index

A

Aerated water, 1

AFDA, 15

AGMARK, 75

Anand, 37

APEDA, 30, 51

Arunachal Pradesh, 7, 43, 53, 57, 58

Asian countries, 1

Assam, 7, 39, 41, 61, 79

B

Balayogi, G.M.C., 57

Banking and finance, 61

Bhutan, 121

Bhutan, 7

Bihar Agricultural Produce Market Act, 77

Brazil, 6, 66

C

Centre for Technology Development, 15

CFTRI, 39

Chilly, 82

China, 3, 7, 121, 135, 136

Chopra, Vipla, 37

Communication, 58

CWC, 77

D

Dehydrated foods, 2

Dept. of Food Processing Industries, 55

District Industries Centres (DICs), 35

F

FCI, 77

FDI, 4

FMCG, 1

Food processing industry, 2
 institutional support to, 28
 packaging strategy, 108
 present status of, 48-62
 progress of in India, 43
 role of govt. in, 10
 selling techniques, 107

Food products,
 export of, 69
 marketing of, 68

FPO, 29, 49, 63

Fruit processing units,
 Govt.'s attitude towards, 112
 performance evaluation of, 93-129

G

Gangtok, 57

Germany, 50, 51

Ginger, 82

Gogoi, 80
Golden rule, 135
Govt. Fruit Preservation Factory, Meghalaya, 94
Guwahati, 34, 37, 38, 132

H

Hibi Fruit Products, 98
Horticulture activities, 78
Horticulture produce,
  status of marketing, 79
IDBI, 23
IEMs, 4
IID, 18
India, 1, 2, 3, 6, 64, 65,
  fruit crops in, 5
Information Technology, 35
  ICAR, 33, 81
ITDP, 15

J

Jackfruit, 54
Joint Ventures, 50
Jorhat, 79

K

Kamjong, 87
Kolkata, 133

L

Lal, Girdhari, 37
Linkages, backward, 14
  forward, 14
Loan facilities, 111

M

MAGFRUIT, 93
Maini, 37
Malaysia, 6
Manipur, 7, 17, 39, 41, 84
  lime and lemon map of, 84
  viability and suitability of horticulture business in, 81
Market legislation,
  enactment of, 76
MATA, 17
Meghalaya, 7, 41, 61, 62, 94
Minis of Food Processing Industries (MFPI), 10, 29, 69
  infrastructure development, 12
  policy, 11
  R&D schemes, 15
  role of, 11
Mizoram, 7, 41, 54, 61
MODVAT, 20
Myanmar, 121, 136

N

NABARD, 32
NAFED, 69
Nagaland, 7, 57, 73
National Horticulture Board, 29
Nayak Committee, 20
NCDC, 31
NEDFC, 19, 34
NEITCO, 38
NERAMAC, 32, 33, 80
NHDC, 22
North East India (Region), 36, 57
  area and population, 8
  basic issues to marketing in, 73
  climate, 7
  development of industrial infrastructure, 18

features of new Industrial policy and other concession in, 18
fiscal incentives to new industries, 19
geographical position, 7
horticulture in, 23
infrastructural development in, 50
list of horticulture based industries, 23
physical features, 7
production and marketing of fruits, 71
technology mission, 23

P

PMRY, relaxation in norms, 21
Power, 58
Present study,
chapter planning, 44
findings and suggestions, 130-138
limitations of, 43
objectives of, 39
significance of, 43
Processed food products, production and marketing of, 63-92
Prusty, S.R., 3

Q

Quality Control, 51
Quality Control Laboratory, 15
Quick frozen food, 2

R

R&D studies, 2
Ratna Fruit India, Manipur, 96
Regional marketing organization, 74
ROCE, 119

S

SANA, 97
SIDBI, 19
Sikkim, 7
Singh, H.P., 6, 37
South Africa, 6
Sunrise Industry, 2

T

Tata Consultancy Services, 37
Technology, 3
Thailand, 134
Thangjam Agro Industries, 97
Trade fairs, 109
Transport Subsidy Scheme, 19
Tripathi, D.P., 3
Tripura, 43, 56, 61

U

UK, 50, 69
Ukhrul, 87
Union Budget, 4
USA, 6, 50, 60

V

Vegetable crops, 90

W

WAIFRUIT, 95, 119
World War I, 49
WSCs, 22
WTO, 15, 121, 133, 137, 138

□□□